FORTSCHRITTE DER CHEMISCHEN FORSCHUNG
TOPICS IN CURRENT CHEMISTRY

Herausgeber:
A. Davison · M. J. S. Dewar
K. Hafner · E. Heilbronner
U. Hofmann · K. Niedenzu
Kl. Schäfer · G. Wittig
Schriftleitung: F. Boschke

Band 15, Heft 2, Juli 1970

NEW RESULTS IN BORON CHEMISTRY

H. D. Johnson, II / S. G. Shore
Lower Boron Hydrides

A. Meller
Boron-Nitrogen
Ring Compounds

K. Niedenzu / C. D. Miller
1,3,2-Diazaboracycloalkanes

G. Heller
Borate und Polyborate

Springer-Verlag Berlin Heidelberg GmbH

15. Band, 2. Heft

Fortschritte der chemischen Forschung
Topics in Current Chemistry

H. D. Johnson, II, and
S. G. Shore — Recent Developments in the Chemistry of the Lower Boron Hydrides 87

A. Meller — Preparative Aspects of Boron-Nitrogen Ring Compounds 146

K. Niedenzu and
C. D. Miller — 1,3,2-Diazaboracycloalkanes 191

G. Heller — Darstellung und Systematisierung von Boraten und Polyboraten 206

Herausgeber:

Prof. Dr. A. Davison — Department of Chemistry, Massachusetts Institute of Technology, Cambridge, MA 02139, USA

Prof. Dr. M. J. S. Dewar — Department of Chemistry, The University of Texas Austin, TX 78712, USA

Prof. Dr. K. Hafner — Institut für Organische Chemie der TH 6100 Darmstadt, Schloßgartenstraße 2

Prof. Dr. E. Heilbronner — Physikalisch-Chemisches Institut der Universität CH-4000 Basel, Klingelbergstraße 80

Prof. Dr. U. Hofmann — Institut für Anorganische Chemie der Universität 6900 Heidelberg 1, Tiergartenstraße

Prof. Dr. K. Niedenzu — University of Kentucky, College of Arts and Sciences Department of Chemistry, Lexington, KY 40506, USA

Prof. Dr. Kl. Schäfer — Institut für Physikalische Chemie der Universität 6900 Heidelberg 1, Tiergartenstraße

Prof. Dr. G. Wittig — Institut für Organische Chemie der Universität 6900 Heidelberg 1, Tiergartenstraße

Schriftleitung:

Dipl.-Chem. F. Boschke — Springer-Verlag, 6900 Heidelberg 1, Postfach 1780

Springer-Verlag — 6900 Heidelberg 1 · Postfach 1780
Telefon (06221) 49101 · Telex 04-61723
1000 Berlin 33 · Heidelberger Platz 3
Telefon (0311) 822001 · Telex 01-83319

Springer-Verlag
New York Inc. — New York, NY 10010 · 175, Fifth Avenue
Telefon 673-2660

ISBN 978-3-540-04821-3   ISBN 978-3-540-36201-2 (eBook)
DOI 10.1007/978-3-540-36201-2

Recent Developments in the Chemistry of the Lower Boron Hydrides

H. D. Johnson, II, and Prof. S. G. Shore

The Ohio State University, Department of Chemistry, Columbus, Ohio 43210, USA

Contents

I.	Introduction	88
II.	Structures and General Chemistry	89
III.	Diborane(6)	93
	A. Preparation	93
	B. Bridge Dissociation Energy	94
	C. Evidence for Factors which Affect the Course of Bridge Cleavage Reactions; the Existence of Singly Hydrogen-Bridged Boranes	94
	D. Borane Adducts	99
	E. Boronium Ions	112
	F. Diborane(6) Derivatives	116
IV.	Triborane(7) Adducts	119
	A. Structure	119
	B. Boron-11 NMR	119
	C. The Preparation and Chemistry of the Dimethyl Ether Adduct of Triborane(7); Diborane(4) Adducts	120
	D. The Octahydrotriborate(−1) Ion	120
V.	Tetraborane(8) Adducts	122
VI.	Tetraborane(10)	123
	A. Preparation	123
	B. The Reaction with $NaBD_4$ in 1,2-Dimethoxyethane	123
	C. Deuterated Derivatives	123
	D. NMR Spectra	124
	E. Mass Spectra	124
	F. Alkyl Derivatives of Tetraborane(10)	125
	G. 2,2′-Bitetraboranyl	126
VII.	Pentaborane(9)	126
	A. The Octahydropentaborate(−1) Ion	126
	B. Halogen Derivatives	131
	C. Adducts	133
	D. Framework Rearrangement	134
	E. NMR	134
	F. Decaborane(16)	135
VIII.	Pentaborane(11)	135
	A. NMR	135
	B. Isotropically Labeled Pentaborane(11)	135
IX.	Hexaborane(10)	136
	A. Preparation	136
	B. Brønsted Acidity	136
	C. Lewis Acidity	137
	D. Hexaborane(10) as a Base	138

X. Hexaborane(12) 138
  A. Preparation and Characterization 138
  B. Chemistry 138

XI. Heptaborane 138

XII. Octaborane(12) 138
  A. Preparation 138
  B. Chemistry 139

XIII. References 139

I. Introduction

This article is concerned with the recent chemistry of the lower boron hydrides; namely

$$B_2H_6, \ B_4H_{10}, \ B_5H_9, \ B_5H_{11},$$

$$B_6H_{10}, \ B_6H_{12}, \ B_8H_{12}, \ B_8H_{14},$$

and their derivatives. Discussion, for the most part, encompasses researches which have been reported after 1964. However, the scope of the presentation does not include carborane derivatives nor borohydride ion derivatives as specific topics. These areas have become so formidable that they are best treated in articles which are devoted to them exclusively[a].

----

[a] Summaries of earlier chemistry, structure, and bonding may be found in the following references. Selected chapters are listed of those volumes which are collections of articles.

i. *Lipscomb, W. N.:* Boron Hydrides. New York: W. A. Benjamin 1963.

ii. *Adams, R. M.:* Boron, Metallo-Boron Compounds and Boranes. New York: Wiley 1964. — Chapter 6: Adams, R. M., Siedle, A. R.: The Hydroboron Ions (Ionic Boron Hydrides). — Chapter 7: Adams, R. M.: The Boranes or Boron Hydrides.

iii. Steinberg, H., McCloskey, A. L.: Progress in Boron Chemistry. New York: Macmillan 1964. — Chapter 2: Coyle, T. D., Stone, F. G. A.: Some Aspects of the Coordination Chemistry of Boron. — Chapter 3: Campbell, G. W., Jr.: The Structures of the Boron Hydrides. — Chapter 10: Schaeffer, R.: Nuclear Magnetic Resonance Spectroscopy of Boron Compounds.

iv. Holtzman, R. T.: Production of Boranes and Related Research. New York: Academic Press 1967.

v. Muetterties, E. L.: The Chemistry of Boron and Its Compounds. New York: Wiley 1967. — Chapter 1: Muetterties, E. L.: General Introduction to Boron Chemistry. — Chapter 5: Hawthorne, M. F.: Boron Hydrides.

vi. Jolly, W. L.: Preparative Inorganic Reactions. New York: Wiley 1968. — Section 2: Parry, R. W., Walter, M. K.: The Boron Hydrides.

vii. Eaton, G. R., Lipscomb, W. N.: NMR Studies of Boron Hydrides and Related Compounds. New York: W. A. Benjamin 1969.

While the chemistry of decaborane(14) has received considerable attention in the last decade because of the varied types of derivatives which can be produced, advances leading to the emergence of significant principles have also occurred in the chemistry of the lower boron hydrides. It now appears as if this area is on the verge of rapid development and expansion.

For the purpose of describing general chemistry, we place the boron hydrides in one of two groups based upon the presence or absence of BH_2 units in the individual structures. Placing hydrides in these two catagories as a function of structure was first suggested by Parry and Edwards [1]. A number of generalizations presented in this chapter represent extensions of principles which were first set forth in this now classic article.

II. Structures and General Chemistry

Fig. 1 depicts molecular structures and 2-dimensional topological representations of the hydrides under consideration. The molecular structure of B_6H_{12} has not yet been positively established; the 2-dimensional topological representation shown is based upon its boron-11 nmr spectrum. Octaborane(14) is not shown in Fig. 1 because of insufficient structural information. However, a possible arrangement is discussed in Section XII.

For the set of hydrides which contain BH_2 units (B_2H_6, B_4H_{10}, B_5H_{11}, and B_6H_{12}), reactions with nucleophiles are intimately associated with the bridge system, giving rise to fragmentation of the molecule. For diborane(6), which is the most extensively studied hydride of this group, two types of fragmentation, so called "symmetrical" and "unsymmetrical" cleavage of the bridge system have been observed:

$+ \; 2 \, L \longrightarrow 2 \, LBH_3$ symmetrical cleavage

where L = nucleophile such as $N(CH_3)_3$

$+ \; 2 \, L \longrightarrow [H_2BL_2]^+BH_4^-$ unsymmetrical cleavage

where L = nucleophile such as NH_3

Relatively few examples of unsymmetrical cleavage are known [2-6].

Analogous formation of molecular and ionic products has been observed for B_4H_{10} [1, 7-11] as well, and is expected for all of the boron hydrides listed in the group containing BH_2 units.

$$+ 2\,L \longrightarrow LBH_3 + LB_3H_7$$

$$+ 2\,L \longrightarrow [H_2BL_2]^+[B_3H_8]^-$$

Factors which determine the course of bridge cleavage have been considered and investigated [2,5,15]. It seems as if reactions with nucleophiles can best be described in terms of a two step sequence in which

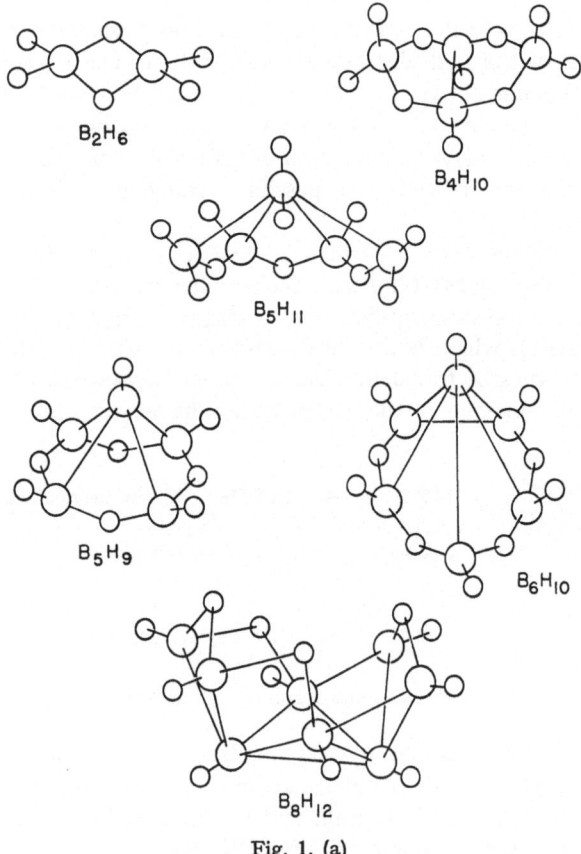

B_2H_6

B_4H_{10}

B_5H_{11}

B_5H_9

B_6H_{10}

B_8H_{12}

Fig. 1. (a)

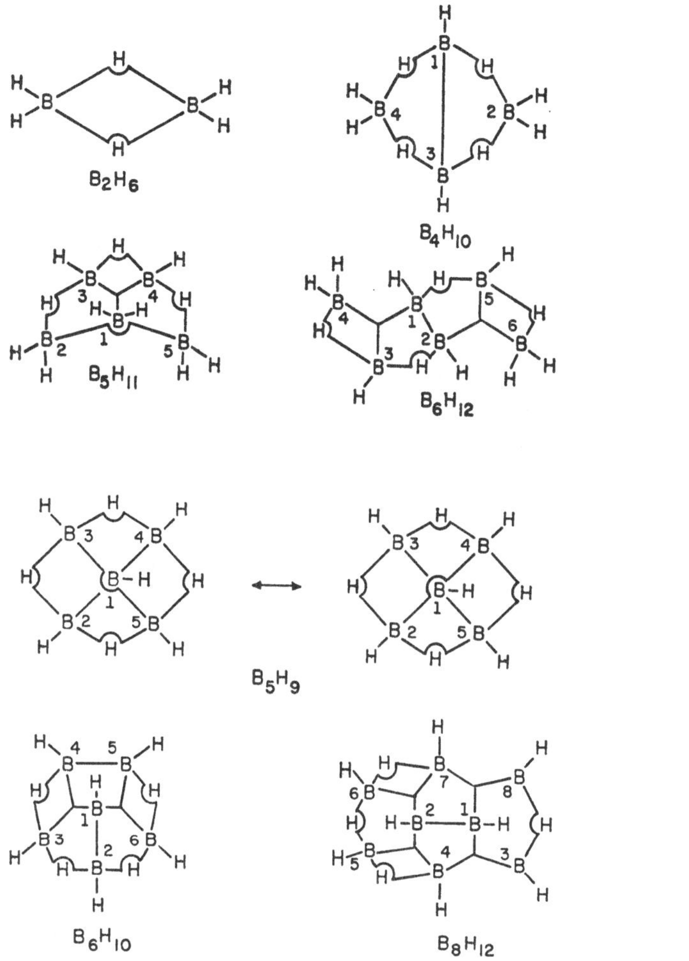

Fig. 1. (b)

Fig. 1. (a) Molecular structures; (b) Two-dimensional topological representations (two resonance forms are necessary for B_5H_9 to preserve its molecular symmetry)

the first step involves displacement of one bridge hydrogen to form a single hydrogen-bridged species:

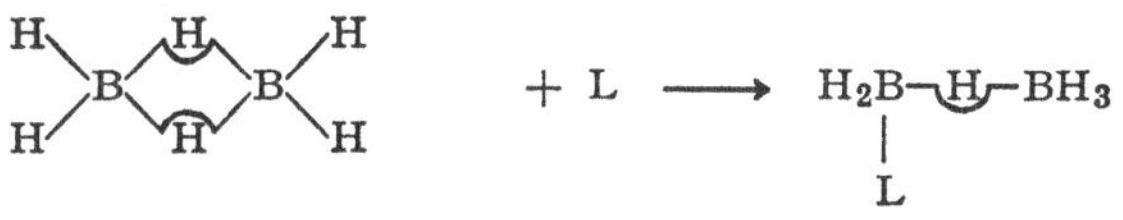

91

The second step is believed to involve displacement of the remaining bridge hydrogen, thereby determining the type of product produced:

$$H_2B{-}H{-}BH_3 + L \longrightarrow 2LBH_3$$
$$\underset{L}{|}$$

or

$$H_2B{-}H{-}BH_3 + L \longrightarrow H_2BL_2{}^+BH_4{}^-$$
$$\underset{L}{|}$$

For the sequence of steps given above, three principal factors have been suggested to determine the course of the reaction [12]. They are: 1) an inductive effect; 2) the inherent donor character of the ligand; 3) a steric effect. With respect to the inductive effect, it has been suggested that in particular with nitrogen bases (since nitrogen is more electron withdrawing than hydrogen) the boron atom which is bonded to nitrogen in the singly-bridged adduct will have a slightly more positive residual charge than the boron which is bound to hydrogens exclusively. Thus in the absence of other factors, it is to be expected that in the second step of the sequence the nucleophile will displace hydrogen from the more positive boron, thereby producing unsymmetrical cleavage.

Concerning the inherent donor character of the ligand, this point simply considers the possibility that while the inductive effect might favor unsymmetrical cleavage, the base L could be a weak donor to boron compared to hydride. Thus formation of the symmetrical cleavage product could occur after initial unsymmetrical cleavage of the bridge system.

$$BH_2L_2^+BH_4^- \longrightarrow 2\,LBH_3$$

Such behavior has been observed in the reaction of tetrahydrofuran with B_4H_{10} [10] and possibly with B_2H_6 [13], but has not been observed with nitrogen bases [2].

The effect of steric factors on the course of bridge cleavage has been established. [12] As the ligand increases in bulk, the tendency for symmetrical cleavage increases. By the same token as substitution on boron increases, the tendency for symmetrical cleavage increases. Both of these observations are consistent with the two-step sequence outlined above in that for the second step of the sequence, nucleophilic displacement of bridge hydrogen would be expected at the less hindered boron atom (that boron from which hydrogen has not been displaced in the first step of the sequence).

The hydrides of the second set shown in Fig. 1 (B_5H_9, B_6H_{10}, B_8H_{12}) do not appear to undergo reactions which are analogs of symmetrical and unsymmetrical cleavage in the sense considered in the preceding discussion. The predominant form of reaction with nucleophiles is either addition to the framework or bridge proton abstraction. Of the two types, the latter reaction has been the better characterized. Thus, strong bases (CH_3^-, H^-) deprotonate B_5H_9 [14,15,16,17] and B_6H_{10} [17,18] at bridge sites to produce the conjugate bases $B_5H_8^-$ and $B_6H_9^-$. The analogous reaction of B_8H_{12} has not yet been established; however, it has been shown that in a series of analogous boron hydrides, the protonic character of the bridge hydrogens increases with increasing size of the boron framework. This has been established by proton competition reactions:

$$B_5H_8^- + B_6H_{10} \longrightarrow B_5H_9 + B_6H_9^-$$
$$B_6H_9^- + B_{10}H_{14} \longrightarrow B_6H_{10} + B_{10}H_{13}^-$$

which show the order of Brønsted acidity to be $B_5H_9 < B_6H_{10} < B_{10}H_{14}$ [18]. Thus, B_8H_{12} is expected to be a stronger Brønsted acid than B_6H_{10}.

Concerning addition reactions, a number of adducts have been reported; however, relatively few products have been characterized definitively. While an overall view of addition is yet to be established, the recently prepared adducts $B_5H_9[P(CH_3)_3]_2$ [19] are tractable with respect to characterization and suggest a possible reaction scheme of these polyhedral frameworks with Lewis bases. The adduct derived from the addition of trimethylphosphine to pentaborane(9) is believed to be geometrically similar to B_5H_{11}, but isoelectronic with $B_5H_{11}^{-2}$. It is suggested that in the reaction of a polyhedral borane such as B_5H_9 or B_6H_{10} with a Lewis base, the first ligand displaces a bridge hydrogen to form a bond with a basal boron. The resulting rearrangement of the framework and electron distribution favors attachment of the second ligand to the apical boron. The trimethylphosphine adducts of B_5H_9 and B_6H_{10} are discussed in greater detail in Sections VII.C. and IX.C., respectively.

The previous discussion has attempted to generalize those aspects of the chemistry of the lower boron hydrides where patterns are apparently beginning to emerge as the chemistry becomes better defined. In the next section we consider the lower boron hydrides on individual bases giving results of recent studies.

III. Diborane(6)

A. Preparation

A number of procedures for the preparation of B_2H_6 have been described recently [21]. From work conducted in this laboratory, we favor the procedure of Long and Freeguard [22] which involves the reaction of

NaBH$_4$ with I$_2$ in diglyme. Not only are the reactants readily available and easy to handle, but the reaction is easily controlled and a pure product is obtained in 98% yield.

B. Bridge Dissociation Energy

Determination of the value of the bridge dissociation energy of diborane(6) has been a continuing problem. Table 1 lists most of the values which have been reported since 1956. The range of energies is 28—59 kcal/mole. They have been obtained from kinetic studies of labile BH$_3$ adducts and from mass spectrometry. An early value for the bridge dissociation energy which was based upon a calorimetric investigation of amine-boranes [23] has since been discounted on the grounds that the reactions studied did not necessarily produce the products claimed [2].

Table 1. *Reported values of the bridge dissociation energy of diborane(6)*

D(kcal/mole)	Method	Year
28.4 ± 2	Kinetic [27]	1956
32—38.3	Kinetic [28]	1964
55 ± 8 [1]	Mass Spec. [25]	1964
39 ± 1	Mass Spec. [24]	1964
37.1 ± 4	Mass Spec. [29]	1965
33.5 ± 2.4	Kinetic [30]	1966
35.0	Kinetic [31]	1966
59.0	Mass Spec. [26]	1967
36 ± 3	Mass Spec. [32]	1969
59.0	Mass Spec. [33]	1969

[1] The authors have subsequently expressed reservations concerning the significance of this value; see Stafford, F. E., Pressley, G. A., Jr., Baylis, A. B.: Advan. Chem. Ser. *72*, 137 (1966). See reference [33] for additional comments.

It is of interest to note that the mass spectrometric studies have led to the detection of BH$_3$ at high temperatures [24–26].

C. Evidence for Factors which Affect the Course of Bridge Cleavage Reactions; the Existence of Singly Hydrogen-Bridged Boranes

In general, reactions of diborane with Lewis bases produce either symmetrical or unsymmetrical cleavage of the bridge system as discussed in

Section II. Most of the known reactions yield symmetrical cleavage products. The classic example of unsymmetrical cleavage is the reaction of diborane with ammonia [1,2].

$$B_2H_6 + 2\,NH_3 \longrightarrow BH_2(NH_3)_2^+BH_4^-$$

Methylamine also produces unsymmetrical cleavage, while dimethylamine produces a mixture of unsymmetrical and symmetrical cleavage products [2]. In addition to these amines, dimethylsulfoxide is the only other base which has been shown to cause unsymmetrical cleavage [3]; $BH_2[OS(CH_3)_2]_2BH_4$.

Substituted diboranes derived from the hydroboration of 1,3-butadiene: 1,2-tetramethylenediborane(6), *1* and 1,2-bis(tetramethylene)-diborane(6), *2* undergo symmetrical and unsymmetrical cleavage reactions [12].

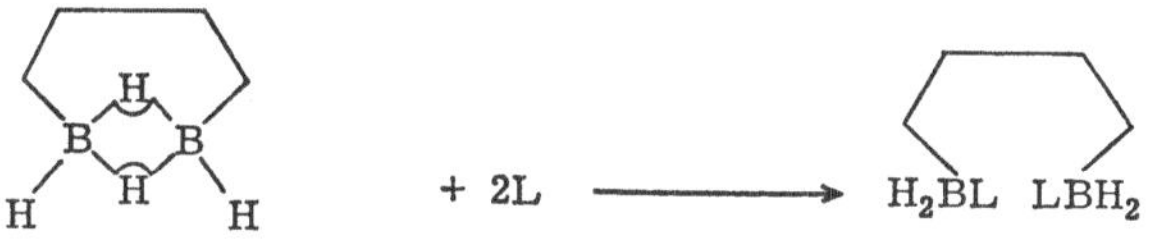

$+ 2L \longrightarrow H_2BL \quad LBH_2$

1

$L = N(CH_3)_3,\ NH(CH_3)_2$

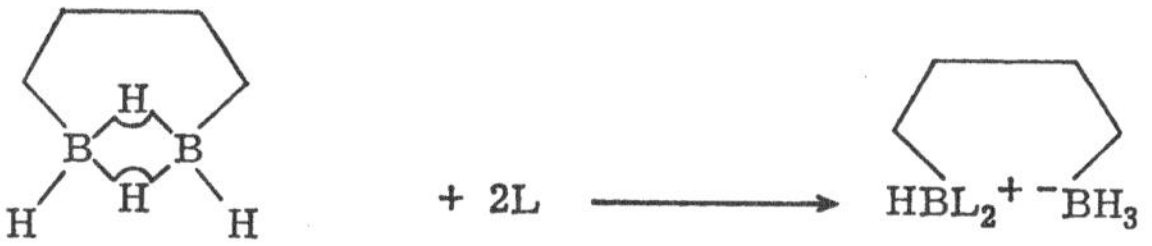

$+ 2L \longrightarrow HBL_2^+ \ ^-BH_3$

1

$L = NH_3,\ NH_2CH_3$

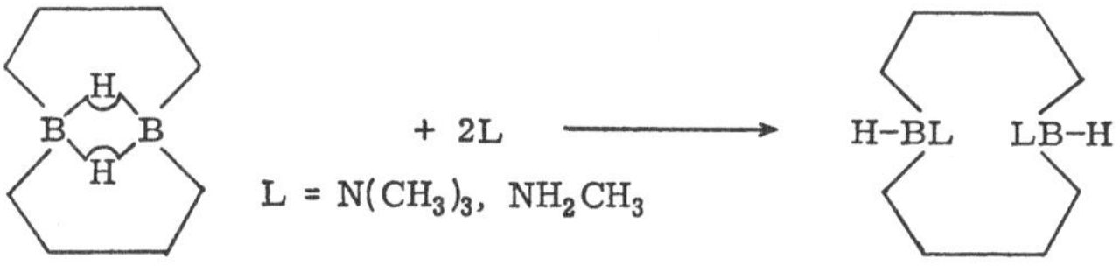

$+ 2L \longrightarrow H-BL \quad LB-H$

$L = N(CH_3)_3,\ NH_2CH_3$

2

$+ 2\,NH_3 \longrightarrow (NH_3)_2B^+ \ ^-BH_2$

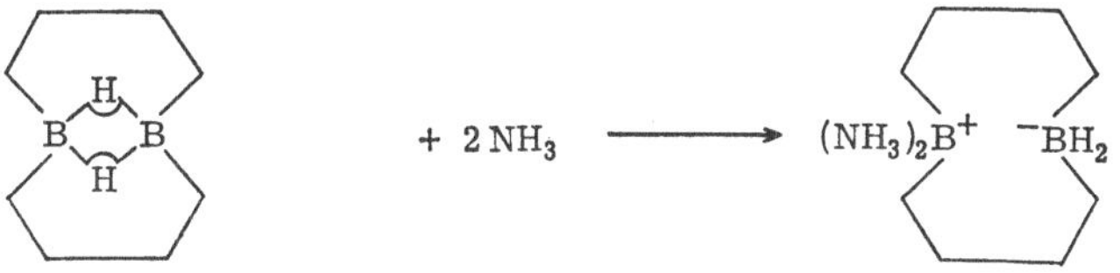

2

The effect of steric factors in determining the course of bridge cleavage is illustrated by the fact that symmetrical cleavage products predominate with increasing methyl substitution from ammonia to trimethylamine. In terms of the two-step reaction sequence given in Section II., it is to be expected that with increasing bulk or steric requirement of the ligand, the tendency for symmetrical cleavage would increase. Furthermore, for a series of diboranes in which terminal hydrogens have been progressively substituted by larger groups, the two-step sequence favors symmetrical cleavage with increasing substitution of the terminal positions of diborane. Information which is presented in Table 2 supports this contention.

There is only indirect evidence for the existence of singly hydrogen-bridged intermediates in the reaction of diborane(6) with Lewis bases [34–36]. However, it has been possible to prepare such materials [37]. The reactions given below produce identifiable materials which are the postulated intermediates in the direct reaction of diborane(6) with ammonia and trimethylamine:

$$\tfrac{1}{2} B_2H_6 + H_3BNH_3 \longrightarrow \underset{\underset{NH_3}{|}}{H_2B\text{--}H\text{--}BH_3}$$

$$\tfrac{1}{2} B_2H_6 + H_3BN(CH_3)_3 \longrightarrow \underset{\underset{N(CH_3)_3}{|}}{H_2B\text{--}H\text{--}BH_3}$$

When these materials are treated with additional Lewis base, the following reactions are observed, giving the same products obtained from the direct reaction of diborane(6) with ammonia and trimethylamine:

$$\underset{\underset{NH_3}{|}}{H_2B\text{--}H\text{--}BH_3} + NH_3 \longrightarrow H_2B(NH_3)_2^+BH_4^-$$

$$\underset{\underset{N(CH_3)_3}{|}}{H_2B\text{--}H\text{--}BH_3} + N(CH_3)_3 \longrightarrow 2\,H_3BN(CH_3)_3$$

The existence of singly hydrogen-bridged structures such as shown above has been questioned and an alternative has been proposed [38].

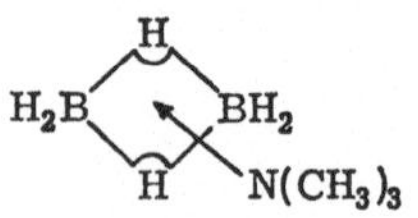

Table 2. *Relative yields of symmetrical and unsymmetrical cleavage products*

Reactants		Unsymmetrical Cleavage Product	Relative Yield	Symmetrical Cleavage Product
B_2H_6	$+\ 2\,NH_3$	$[BH_2(NH_3)_2^+][BH_4^-]$		No evidence for product
	$+\ 2\,NH_2CH_3$	$[BH_2(NH_2CH_3)_2^+][BH_4^-]$	$\gg$	$CH_3NH_2BH_3$
	$+\ 2\,NH(CH_3)_2$	$[BH_2(NH(CH_3)_2)_2^+][BH_4^-]$	$<$	$(CH_3)_2NHBH_3$
	$+\ 2\,N(CH_3)_3$	No evidence for product		$(CH_3)_3NBH_3$
(diborane structure with B–H–B bridges)	$+\ 2\,NH_3$	$(H_3N)_2BH^+\ {}^-BH_3$		No evidence for product
	$+\ 2\,NH_2CH_3$	$(NH_2CH_3)_2BH^+\ {}^-BH_3$		No evidence for product
	$+\ 2\,NH(CH_3)_2$	No evidence for product		$(CH_3)_2NHBH_2 \quad BH_2NH(CH_3)_2$
	$+\ 2\,N(CH_3)_3$	No evidence for product		$(CH_3)_3NBH_2 \quad BH_2N(CH_3)_3$
(bridged boron hydride structure, B–H–B)	$+\ 2\,NH_3$	$(NH_3)_2B^+\ {}^-BH_2$		No evidence for product
	$+\ 2\,NH_2CH_3$	No evidence for product		$CH_3NH_2BH \quad BHNH_2CH_3$
	$+\ 2\,NH(CH_3)_2$	No evidence for product		$(CH_3)_2NHBH \quad BHNH(CH_3)_2$
	$+\ 2\,N(CH_3)_3$	No evidence for product		$(CH_3)_3NBH \quad BHN(CH_3)_3$

However, available experimental facts do not lend credence to this structure [34].

Although several singly hydrogen-bridged boranes have been reported [39-44], with the exception of $(C_2H_5)_4NB_2H_7$ [44] and $(C_4H_9)_4$ NB_2H_7 [42] they cannot be isolated as stable materials.

Tetraethylammonium diborhydride has been prepared from the following reaction:

$$B_2H_6 + 2\ (C_2H_5)_4NBH_{4(s)} \xrightarrow{-78°} 2\ (C_2H_5)_4N[H_3B\text{—}H\text{—}BH_3]_{(s)}$$

From the heat of this reaction, the dissociation energy of B_2H_6, and other considerations, the single hydrogen-bridge dissociation energy of $H_3B\text{—}H\text{—}BH_3^-$ has been estimated [44]. The value ranges from 31 ± 8 kcal/mole to 43 ± 8 kcal/mole depending upon the choice of dissociation energy of B_2H_6 (35 kcal/mole or 59 kcal/mole).

A transannular singly hydrogen-bridged ring structure has been prepared through the following reaction.

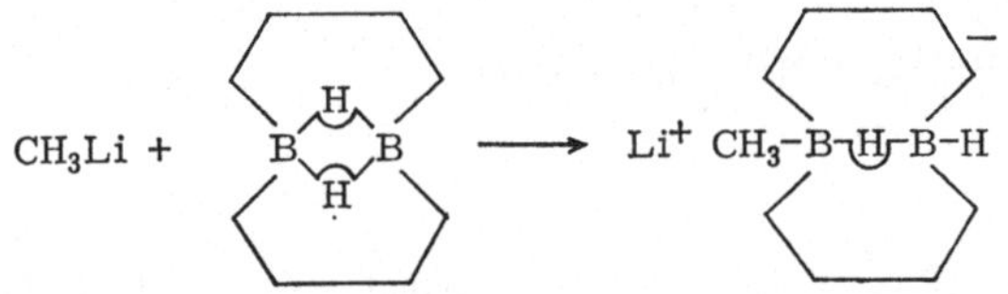

While this material has not been isolated, it is stable in ether solution. It boron-11 nmr spectrum consists of a singlet and doublet in the area ratio of 1:1 which is consistent with the proposed structure; in general, bridge coupling is not observed for singly hydrogen-bridged structures.

The structure depicted above, 3, is the isoelectronic analog of the singly hydrogen-bridged intermediate expected in the reaction of 2 with ammonia. When this intermediate is treated with ammonia, the analog of unsymmetrical cleavage is produced.

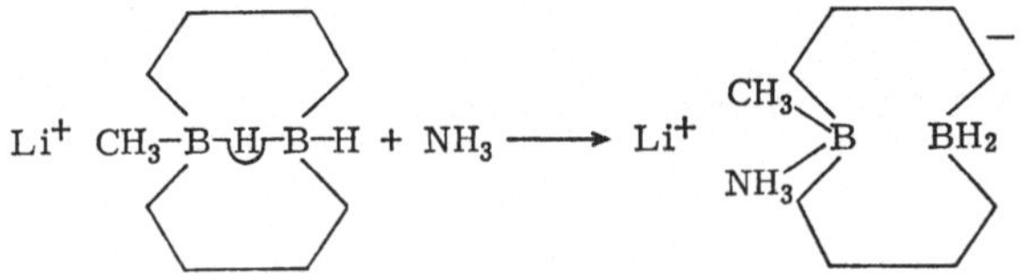

The boron-11 nmr spectrum of the product consists of a singlet and a triplet in the area ratio of 1:1. When 3 is treated with trimethylamine, the analog of symmetrical cleavage is produced.

$$\text{Li}^+ \quad CH_3\text{-B-H-B-H} + N(CH_3)_3 \longrightarrow \text{Li}^+ \quad \begin{array}{c} H \\ B \\ CH_3 \end{array} \begin{array}{c} H \\ B \\ N(CH_3)_3 \end{array}$$

3

The boron-11 nmr spectrum of the product consists of two doublets in the area ratio 1:1.

D. Borane Adducts

Since 1964, a large number of borane adducts have been prepared through either the direct reaction of diborane(6) with Lewis bases or through less direct procedures which involve displacement of hydride ion from BH_4^-. Table 3 below lists most of these materials and some of their properties.

In the following sub-sections we discuss a relatively small number of compounds which, to us, are among the more unusual borane adducts investigated since 1964.

a) Borane Analogs of Oxyanions

A number of years ago, Mullikan [83] noted that BH_3 is isoelectronic with oxygen. Although this analogy is subject to criticism on theoretical grounds, it has proved to be a fruitful starting point for Parry and co-workers who developed a system of borane chemistry in which BH_3 adducts that are isoelectronic analogs of oxyanions have been prepared. The first application of this analogy led to the determination of the structures of the bisamine adducts of borane-carbonyl [47,48]. By considering BH_3CO to be isoelectronic with CO_2, Carter and Parry [47,48] reasoned that the bisammonia adduct is structurally analogous to ammonium carbamate, the product obtained from the reaction of carbon dioxide with ammonia.

$$BH_3CO + 2\,NH_3 \longrightarrow NH_4^+ \; H_2N\overset{\displaystyle O}{\overset{\|}{-C}}-BH_3^-$$

Adducts with methylamine and dimethylamine have also been prepared [47,48]. All of the adducts are stable, crystalline solids which were shown to have the carbamate structure from a variety of observations. Trimethylamine does not react with BH_3CO to give a carbamate analog; an unstable 1:1 adduct is obtained.

Table 3. *Borane Adducts*

Adduct	Spectra Reported	Comments and Physical Properties	References
$\left[H_3B\!-\!CO \right]^{-2}$ (O=)	^{1}H and ^{11}B nmr, IR, Powder pattern		46)
$\left[H_3B\!-\!C\!-\!NR_2 \right]^{-}$ (O)	IR, Powder pattern		47,48)
H_3B / N(H)(CH$_2$)(CH$_2$) ring	^{1}H and ^{11}B nmr, IR	m. 47—48°	49)
H_3B / N(H)(CHCH$_2$NH$_2$BH$_3$)(CH$_2$) ring	^{1}H and ^{11}B nmr, IR		49)
$(CH_3)_2$ $H_3B\!-\!N$—CH$_2$ / $H_3B\!-\!N$—CH$_2$ $(CH_3)_2$		sublimes at 70°	50)

Compound	Methods	Notes	Ref.
$H_3B\text{–}N(CH_3)_2C_2H_4N(CH_3)_2$	^1H and ^{11}B nmr, IR	disproportionates at room temperature	51)
$\underset{H_3B}{\overset{CH_3}{>}}N\underset{CH_2\text{–}CH_2}{\overset{CH_2\text{–}CH_2}{}}N\underset{CH_3}{\overset{CH_3}{<}}$	^1H and ^{11}B nmr, IR		52)
$\underset{H_3B}{\overset{CH_3}{>}}N\underset{CH_2\text{–}CH_2}{\overset{CH_2\text{–}CH_2}{}}N\underset{BH_3}{\overset{CH_3}{<}}$	^1H and ^{11}B nmr, IR		52)
$H_3B\text{–}N\underset{CH_2\text{–}CH_2}{\overset{CH_2\text{–}CH_2}{\diagdown}}\overset{CH_2\text{–}CH_2}{}N$	^1H and ^{11}B nmr, IR	Stable at room temperature	52)
$[H_3BN(CH_3)_2]^-$	^1H and ^{11}B nmr, IR		224)
$[(H_3B)_2N(CH_3)_2]^-$			225)
$H_3B\text{–}N\underset{CH_2\text{–}CH_2}{\overset{CH_2\text{–}CH_2}{\diagdown}}\overset{CH_2\text{–}CH_2}{}N\text{–}BH_3$	^1H and ^{11}B nmr, IR		52)

Table 3 (continued)

Adduct	Spectra Reported	Comments and Physical Properties	References
$H_3B-N(CH_3)NHBH_2$	IR		53)
$H_3B-\overset{H}{\underset{\vert}{N}}=C(C_6H_5)_2$	IR	decomposes slowly at 70°	54)
$H_3B-N(CH_3)CH_2BH_2-N(CH_3)_3$	IR	sublimes at 100° in high vacuum	55)
$H_3B-NHC\overset{O}{\overset{\Vert}{C}}NH(C_6H_5)$, CH_3	IR	m. 142—146°	56)
$H_3B-NHC\overset{O}{\overset{\Vert}{C}}NH(C_6H_5)$, C_2H_5	IR	m. 152°	56)
$H_3B-NC\overset{O}{\overset{\Vert}{C}}NH(C_6H_5)$, $(C_2H_5)_2$	IR	m. 122—124°	56)

$\overset{O}{\underset{\underset{C(CH_3)_3}{\vert}}{\parallel}}$ $H_3B-NHCN(C_6H_5)$	IR	m. 170—175°	56)
$\overset{O}{\underset{\underset{C(CH_3)_3}{\vert}}{\parallel}}$ $H_3B-NHCNH(C_2H_5)$	IR	decomposes without melting	56)
$(CH_3)_2NC_2H_4NC_2H_4N(CH_3)_2$ with CH_3, BH_3, BH_3, BH_3	IR	m. 185—186°	57)
$(CH_3)_2NC_2H_4NC_2H_4NC_2H_4N(CH_3)_2$ with CH_3, CH_3, BH_3, BH_3, BH_3	IR	m. 118.5—119.5°	57)

103

Table 3 (continued)

Adduct	Spectra Reported	Comments and Physical Properties	References
$(CH_3)_2NC_2H_4NC_2H_4NC_2H_4N(CH_3)_2$ — with BH_3 BH_3 BH_3 BH_3	IR	m. 208–210°	57)
$(CH_3)_2NC_3H_7NC_3H_7N(CH_3)_2$ — with CH_3, BH_3 BH_3 BH_3	IR	m. 185–186.5°	57)
$(CH_3)_2$ — $H_3B{-}N{-}SiH_2$ — $H_3B{-}N{-}$ $(CH_3)_2$		dissociates to the mono-adduct on distillation from −84°	50)
$(CH_3)_2$ — $H_3B{-}N{-}SiH_2$ — $N{-}$ $(CH_3)_2$		m. 10°	50)

Compound	Characterization	Notes	Ref.
H_3B-N $\begin{matrix}(CH_3)_2\\ \diagdown\\ \diagup\end{matrix}$ $SiHN(CH_3)_2$ H_3B-N $(CH_3)_2$			50)
$H_3B-NSi[N(CH_3)_2]_3$ $(CH_3)_2$		dissociates on pumping	50)
H_3B-PF_2H	^1H ^{11}B and ^{19}F nmr, IR mass spectrum	b. p 6.2° (extrapolated); not measurably dissociated at 1 atm. and 25°.	58)
$H_3B-P[N(CH_3)_2]_xF_{3-x}$	IR	$x = 1,2$	59)
$H_3B-P[NH(CH_3)]_xF_{3-x}$	IR	$x = 1, 2, 3$	59)
$H_3B-PF_2PF_2$	^{11}B nmr, IR, mass spectrum	decomposes slowly at ambient temperature to $BH_3PF_3 + (PF)_x$	60)
$(BH_3)_n \cdot P[N(CH_3)CH_2]_3CCH_3$	^1H nmr	$n = 1$ (PB) and 2	61)

Table 3 (continued)

Adduct	Spectra Reported	Comments and Physical Properties	References
$H_3B \cdot XP[N(CH_3)CH_2]_3CCH_3$	^1H nmr	$x = 0, S$	61)
$H_3B-P(C_2H_5)_2Si(CH_3)_3$	IR	m. 12°	62)
$H_3P-P(C_4H_9)_2Si(CH_3)_3$	IR	m. -18 to $-19°$	62)
$H_3B-PH_2GeH_3$	^1H nmr		63)
$H_3B-PH_2SiH_3$	^1H nmr	decomposes slowly at room temp; completely dissociated in the gas phase	64)
$H_3B-PH_2Si_2H_5$	^1H	decomposes at 0° to SiH_4, Si_2H_6, PH_3 and B_2H_6	65)
$H_3B-P(CH_3)_2N(CH_3)_2$	^1H, ^{11}B and ^{31}P nmr		66)
$H_3B-P[N(CH_3)_2]_3$	^1H, ^{11}B and ^{31}P nmr		66)
$H_3-P(CH_3)[N(CH_3)_2][OCH(CH_3)_2]$	^1H, ^{11}B and ^{31}P nmr		67)
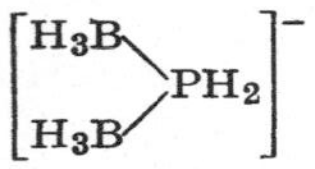 $\left[\begin{smallmatrix} H_3B \\ H_3B \end{smallmatrix} PH_2\right]^-$	^1H, ^{11}B ^{31}P nmr, IR, Raman	Alkylammonium, Na+ and K+ salts air stable; soluble in alcohol, acetone, diglyme, H2O; insoluble in CCl4 and ether	68,69,70,71)

Compound	Method	Notes	Ref.
$H_3B-PF_2OPF_2$	^1H, ^{19}F and ^{11}B nmr	mono-adduct only	226)
$H_3B-P[N(CH_3)N(CH_3)]_3P-BH_3$		m. 250° (dec.)	72)
$(CH_3)_2PCH_2N(CH_3)_2$ with BH_3	^1H nmr		73)
$(CH_3)_2PCH_2N(CH_3)_2$ with BH_3 BH_3	^1H nmr		73)
$(CH_3)_2PCH_2SCH_3$ with BH_3	^1H nmr		73)
$(CH_3)_2NCH_2SCH_3$ with BH_3	^1H nmr		73)
$[(CH_3)_3PCH_2N(CH_3)_2]^+$ with BH_3	^1H nmr		73)
$(CH_3)HPC_3H_5$ with BH_3	IR		74)

107

108

Table 3 (continued)

Adduct	Spectra Reported	Comments and Physical Properties	References
BH_3 BH_3 $\mid$ $(CH_3)HPC_3H_6N(CH_3)_2$			74)
$(H_3B)_n \cdot P_4O_6$	^1H, ^{11}B and ^{31}P nmr, Powder pattern (for n = 2)	n = 1, 2, 3, 4 phosphorus donor	75,76)
$(H_3B)_n \cdot P_4[N(CH_3)]_6$	^1H, ^{11}B and ^{31}P nmr	n = 1, 2, 3, 4 phosphorus donor	77)
$(H_3B)_2 \cdot [As(CH_3)_2]_2PCF_3$	^{11}B nmr	arsenic donor	78)
H_3BSH^-	^{11}B nmr		108)
$(H_3B)_2SH^-$	^{11}B nmr		106,107,108)
$H_3BRe(CO)_5^{\bar{}}$	IR, electronic spectrum		85)
$(H_3B)_2Re(CO)_5^{\bar{}}$	IR, electronic spectrum		85)
$H_3BMn(CO)_5^{\bar{}}$	IR, electronic spectrum	dissociates on standing	85)
$H_3BMn(CO)_4P(C_6H_5)_3^{\bar{}}$	IR, electronic spectrum		85)

The isoelectronic analogy has been extended by Malone and Parry [46] in the following reaction which produces the analog of potassium carbonate.

$$BH_3CO + 2 KOH \longrightarrow K_2H_3BCO_2 + H_2O$$

This salt is a stable, crystalline solid at room temperature. Borane-carbonyl is regenerated by treating the carbonate analog with polyphosphoric acid [46].

$$K_2H_3BCO_2 + 2 H_3PO_4 \xrightarrow{85\% H_3PO_4} 2 KH_2PO_4 + BH_3CO + H_2O$$

Analogs of hypophosphite salts have been prepared by several research groups [68–71].

$$2 BH_3PH_3 + NH_3 \longrightarrow PH_3 + NH_4^+PH_2(BH_3)_2^-$$

Products containing the $(CH_3)_2NH_2^+$ and $(CH_3)NH_3^+$ ions have been prepared by employing the appropriate amine [68]. These quaternary ammonium salts are air stable, crystalline solids. In metal ammonia solutions, the ammonium salt has been converted to sodium and potassium salts [68].

$$[NH_4]^+[PH_2(BH_3)_2]^- + M \longrightarrow MPH_2(BH_3)_2 + \tfrac{1}{2} H_2 + NH_3$$

M = Na, K

Additional preparations of hypophosphite analogs are indicated in the reactions which are given below.

$$KPH_2 + B_2H_6 \xrightarrow[\text{(C}_4\text{H}_9)_2\text{O}]{-78\,°} KPH_2(BH_3)_2 \ [69]$$

$$PH_4I + 2 NaBH_4 \longrightarrow NaPH_2(BH_3)_2 + 2 H_2 + NaI \ [70,71]$$

In Section III.F, the chemistry of $H_3BSBH_3^-$, the analog of the sulfonate ion, is discussed.

b) Phosphine-Borane Adducts

Gamble and Gilmont [82], in 1940, prepared a bisphosphine adduct of diborane(6) which for a number of years was presumed to be PH_4^+ $BH_3PH_2BH_3^-$. Rudolph, Parry, and Farran [79], and McGandy and Eriks [80] have shown that the material is in fact H_3PBH_3 and that it is thermally stable at room temperature.

Fluorophosphine-boranes have been prepared and the relative donor abilities of the phosphine ligands have been reported [58].

$$PF_2H > PF_3 > PH_3 \text{ as a base to } BH_3$$

From gas phase displacement reactions, Rudolph and Parry [58] obtained an equilibrium constant for the reaction.

$$PH_{3(g)} + H_3BPF_{3(g)} \rightleftharpoons H_3BPH_3(s) + PF_{3(g)}$$

$K_{eq} = 2.0 \pm 0.3 \times 10^{-5} \text{ (atm.}^{-1})$

We have recalculated this constant from the published data and obtain a value of $K_{eq} = 2 \times 10^1$ (atm^{-1}). Although this recalculated value indicates that equilibrium is to the right in the equation given above, it has been pointed out that this is not necessarily a result of greater donor ability of PH_3 *versus* PF_3, but is more likely a result of the high lattice energy of $H_3BPH_3(s)$ [58]. If a suitable solvent could be found, it would be of interest to study this system in a homogeneous environment.

Sawodny and Goubeau [81] have examined the vibrational spectra of H_3BPH_3 and H_3BPF_3 and have estimated the respective bond orders to be 0.78 and 0.92.

Diborane(6) reacts with tetrafluorodiphosphine to give only a mono-borane adduct [60].

$$P_2F_{4(g)} + \tfrac{1}{2} B_2H_{6(g)} \longrightarrow P_2F_4BH_{3(g)}$$

This adduct decomposes slowly in the gas phase to form H_3BPF_3 and polymers of empirical composition (PF)x.

c) Borane Adducts to Phosphorus(III) Oxide

A series of adducts of BH_3 to P_4O_6 have been prepared [75] by bubbling B_2H_6 through chloroform solutions of P_4O_6 at room temperature and atmospheric pressure. Gross compositions of up to 2.35 BH_3/P_4O_6 have been obtained. The adducts are labile with rapid equilibration occurring among all of the possible moieties.

$$P_4O_6(BH_3)n$$

n = 1, 2, 3, 4

Crystalline compounds, $P_4O_6(BH_3)_2$ and $P_4O_6(BH_3)_3$, have been isolated from this system. The bis-adduct has also been prepared through the direct combination of reactants in the absence of a solvent [76].

The system $P_4(NCH_3)_6$-B_2H_6 produces a series of adducts analogous to those discussed above [77].

d) A Borane Adduct of the Germyl Hydride Anion

The Lewis basicity of GeH_3^- has been demonstrated in the following reaction [84].

$$KGeH_3 + \tfrac{1}{2} B_2H_6 \longrightarrow KGeH_3BH_3$$

The product is a crystalline solid which melts at about 98° and decomposes at 200° to form germanium hydrides, potassium borohydride, hydrogen, and germanium. In the presence of diborane(6), the following reaction is observed [84].

$$2 \, KGeH_3BH_3 \xrightarrow[\text{1,2-dimethoxyethane}]{B_2H_6} KBH_4 + KGe_2H_5BH_3$$

e) Borane Adducts of Transition Metals

Several examples of borane complexes of transition metals have been reported [b] in which the metal is an apparent Lewis base [85].

The order of basicity of carbonylate ions to BH_3 appears to be

$$Re(CO)_5^- > Mn(CO)_5^- > Co(CO)_4^-$$

The rhenium pentacarbonylate ion will complex with up to two borane units

$$Re(CO)_5^- + THF \cdot BH_3 \longrightarrow H_3BRe(CO)_5^- + THF$$

$$H_3BRe(CO)_5^- + THF \cdot BH_3 \longrightarrow (H_3B)_2Re(CO)_5^- + THF$$

Salts containing the cations Na^+, $[(C_2H_5)_4N]^+$, and $[(C_4H_9)_4P]^+$ have been prepared. They are all stable crystalline solids at room temperature. In ether solution they behave as mild reducing agents.

The structures of these adducts have yet to be established in a definitive manner. However, the infrared spectrum of $[H_3BRe(CO)_5]^-$ is very similar to that of the parent anion in the CO stretching region,

---

[b] While this discussion is concerned exclusively with BH_3 adducts of transition metals, boron halide adducts have been also reported. Scott, R. N., Shriver, D. F., Vaska, L.: J. Am. Chem. Soc. **90**, 1079 (1968), and references therein, Powell, P., Nöth, H.: Chem. Commun. 637 (1966) and references therein.

leading to the conclusion that the most likely structure is the one shown below.

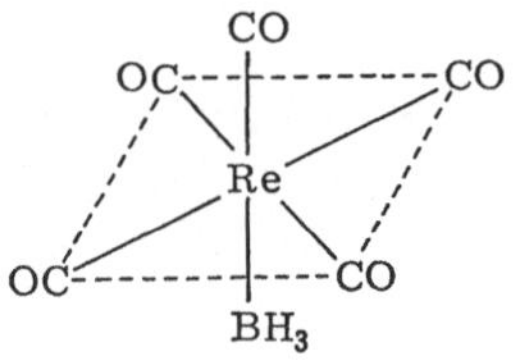

Another possible structure is one in which boron is coordinated to the oxygen of carbonyl group.

In the case of the bisborane adduct, the possible structure which has been proposed by Parshall [85] is a singly hydrogen-bridged borane.

$$H_3B\text{—}H\text{—}BH_2Re(CO)_5]^-$$

However, available spectra (IR, ^1H and ^{11}B nmr) are inconclusive. A second structure considered is one in which two BH_3 groups are bound to seven coordinate rhenium. A third possibility which has been suggested, places one BH_3 on the metal and one on a carbonyl oxygen.

The monoborane adduct of manganese pentacarbonylate ion is significantly less stable than its rhenium analog [85]. At $-25°$ $(C_2H_5)_4N$ $[H_3BMn(CO)_5]$ evolves B_2H_6. The adduct $(C_2H_5)_4N[H_3BMn(CO)_4P(C_6H_5)_3]$, however, is apparently stable at room temperature. This enhanced stability is attributed to $P(C_6H_5)_3$ behaving primarily as a sigma donor, thereby tending to increase the electron density and subsequent donor ability of the metal atom.

The ion $Co(CO)_4^-$ appears to react reversibly with $THF \cdot BH_3$ [85]. Attempts to isolate the sodium salt of the adduct were unsuccessful.

The compound $HMn_3(CO)_{10}(BH_3)_2$ is considered in Section IV. D; it differs appreciably in boron arrangement from the borane adducts discussed above [86].

It is of interest to note that the adduct $Fe(phen)_2(CN)_2(BH_3)_2$ (phen = 1,10-phenanthroline) has been prepared [87]. Cyanide bridges exist between the transition metal and BH_3 groups.

E. Boronium Ions

Direct syntheses of boronium ions can be achieved through unsymmetrical cleavage reactions of B_2H_6 [1-6] or B_4H_{10} [7-11], but is restricted to a relatively small number of ligands. However, through less direct methods, several research groups [88-94] have developed syntheses of boronium ions containing a wide variety of ligands. The equations given below summarize preparative reactions employed. The first three reactions are among the most convenient for preparing bisamine boronium ions.

$$NaBH_4 + 2\,L + 2\,I_2 \xrightarrow{\text{excess L}} H_2BL_2^+I^- + 2\,LH^+I^- + NaI$$

$$LBH_3 + 2\,L + I_2 \xrightarrow{\text{excess L}} H_2BL_2^+I^- + LH^+I^-$$

$$LBH_2R + 2\,L + I_2 \xrightarrow{\text{excess L}} HRBL_2^+I^- + LH^+I^-$$

$L = N(CH_3)_3$, pyridine, and methyl pyridines
$R = C_6H_5$, cyclohexyl

$$LBH_3 + LH^+X^- \xrightarrow{100°-180°} [H_2BL_2]^+X^- + H_2$$

X = large anion such as I^- or $[B_{12}H_{12}]^{-2}$
L = diamines, amines, phosphines

In the preceding reaction, the anion X^- should be reasonably stable toward reduction and be a weak base. Iodide and polyhedral borane ions meet these requirements, but chloride is sufficiently basic to compete with the donor ligand, producing the following reaction as the predominant one.

$$LBH_3 + LH^+Cl^- \longrightarrow LBH_2Cl + H_2 + L$$

Of the ligands given above, diamines produce cyclic cations. When arsine or sulfide are the donor ligands, the reaction to be employed is the following one.

$$LBH_3 + L + HI \longrightarrow [H_2BL_2]^+I^- + H_2$$

Displacement of a ligand from a boronium ion by a more basic ligand is also a useful synthetic procedure.

$$H_2BL_2^+ + 2\,L' \longrightarrow [H_2BL_2']^+ + 2\,L$$

Ability to displace one ligand by another ligand is given by the following order.

diamines > amines > phosphines ~ arsines > sulfides

However, steric factors appear to play an important role in displacement reactions. Thus, for example, a sterically hindered amine base will not displace an unhindered potentially weaker base such as sulfide.

113

Hydrolytic and oxidative stability of the cations increase with the base strength of the ligand. Bistrimethylamine-boronium cations are unusually stable. They have been recovered, apparently unchanged after prolonged periods of time at 100° in concentrated acids (HCl, H_2SO_4, and HNO_3) and 10% NaOH. On the other hand bisdialkyl sulfide boronium ions are rapidly attacked by cold water. The ion $[BH_2(OH_2)_2]^+$ is perhaps the most unstable thermally and hydrolytically of the boronium ions reported. Evidence has been claimed for its existence in 8 MHCl solution at $-70°$ [95,96].

Although boronium salts are most usually formed as iodides, other anions have been incorporated through metathesis reactions and ion exchange procedures in appropriate solvents. Examples of anions used are listed in Table 4.

Table 4. *Anions of boronium salts*

Anions	Comments
$AuCl_4^-$	1, 2)
Br_3^-	2)
$B_{10}H_{10}^{-2}$	2)
$B_{12}H_{12}^{-2}$	2)
$Cr(NH_3)_2(SCN)_4^-$	2)
$(NC)_2CHCH(CN)_2$	2)
$B(C_6H_5)_4^-$	2)
SbF_6^-	2)
F^-	1, 3)
Cl^-	1, 3)
I^-	1, 3, 4, 5, 6)
NO_3^-	3)
SO_4^{-2}	1, 3)
PF_6^-	1, 2, 7, 8)

1) Soluble in CH_2Cl_2 and $CHCl_3$.
2) Slightly soluble in H_2O.
3) Soluble in H_2O.
4) Soluble in acetone.
5) Slightly soluble in CCl_4, ether, and benzene.
6) Difficult to separate from NH_4I in H_2O.
7) Especially good for separation from NH_4^+ salts in H_2O.
8) Soluble in CH_3CN.

Pyrolysis of certain boronium salts leads to the anion serving as a ligand to boron [97].

$$H_2BL_2^+X^- \longrightarrow H_2BXL + L$$

L = N(CH₃)₃	N(CH₃)₃	N(CH₃)₃	N(CH₃)₃
$L = N(CH_3)_3$	$N(CH_3)_3$	$N(CH_3)_3$	$N(CH_3)_3$
$X = N_3^-$	Cl^-	Br^-	$CO_3{}^-_2$
$t = 120-200°$	$200-250°$	$200-250°$	$200°$

Nöth and Beyer [98] report the preparation of $H_2BClN(CH_3)_3$ and $H_2BrBN(CH_3)_3$ through the reaction of trimethylamine-borane with the appropriate hydrogen halide. However, it has been claimed [97] that this reaction yields products which are difficult to purify.

Boronium ions containing mixed neutral ligands have been prepared also. A general preparative reaction is listed below [92,99-102]

$$BH_2IL + L' \longrightarrow BH_2LL'^+ + I^-$$

L and L': NH_3, $N(CH_3)$, t-BuNH₂, $NH(C_2H_5)_2$, pyridine, substituted pyridines, quinoline, N,N-dimethylaniline, dimethylformamide, $(CH_3)_2$-NCH_2SCH_3, $(CH_3)_2PCH_2SCH_3$ pyridine N-oxide, $(CH_3)_3NO$, $(CH_3)_3PO$, $(CH_3)_2SO$, $P(CH_3)_3$, $P(C_2H_5)_3$, $P(C_6H_5)_2NC_2H_4N(C_2H_5)_2$. For ease of handling, mixed ligand boronium salts are generally converted to PF_6^- salts.

Among the more novel examples of boronium cations are homomorphs of norbornane [103]

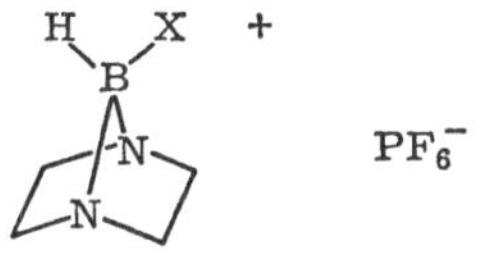

and an asymmetric boronium ion which has been resolved by chemical means [104].

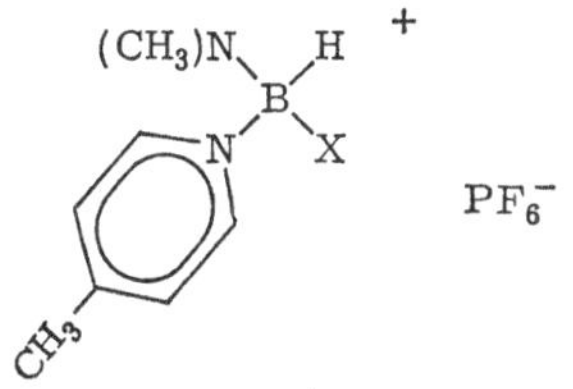

X = Cl, Br

115

Cations containing two and three BH_2 units have been prepared according to the following reactions.

$$(CH_3)_3NBH_2SCH_3 + (CH_3)_3NBH_2I \xrightarrow{CHCl_3} \left[(CH_3)_3NBH_2-\overset{\overset{\textstyle CH_3}{|}}{S}-BH_2N(CH_3)_3 \right]^+ I^-$$

$$(CH_3)_3NBH_2-\overset{\overset{\textstyle CH_3}{|}}{S}-BH_3 \xrightarrow[\text{2) } (CH_3)_3NBH_2SCH_3]{\text{1) } \frac{1}{2} I_2} \left[(CH_3)_3NBH_2-\overset{\overset{\textstyle CH_3}{|}}{S}-BH_2-\overset{\overset{\textstyle CH_3}{|}}{S}-BH_2-N(CH_3)_3 \right]^+ I^- + \frac{1}{2}$$

The iodides are readily converted to PF_6^- salts which are stable in air. The bisboron cation is significantly more stable with respect to hydrolysis than the trisboron cation.

F. Diborane(6) Derivatives

a) μ-Mercaptodiborane(6)

The structure of the $[HS(BH_3)_2]^-$ ion, which can be prepared as shown below, has been established by Keller [107,108].

$$(C_2H_5)_4NSH + B_2H_6 \xrightarrow[-78°]{H_2S(l)} [(C_2H_5)_4N][HS(BH_3)_2] \text{ [106,107]}$$

or

$$(C_2H_5)NBH_4 + H_2S_{(l)} \xrightarrow[-H_2]{-78°} [(C_2H_5)_4N][HSBH_3] \xrightarrow{\frac{1}{2}B_2H_6} [(C_2H_5)_4N][HS(BH_3)_2] \text{ [107,108]}$$

Treatment of solid $[(C_2H_5)_4N][HS(BH_3)_2]$ with anhydrous hydrogen chloride at $-78°$ produces hydrogen and μ-mercaptodiborane(6) in 20—25 percent yields [108-109].

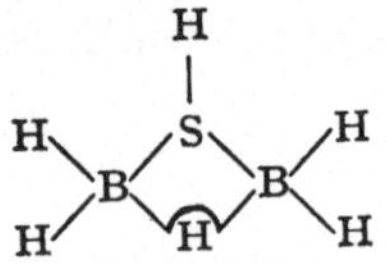

The compound is thermally unstable at room temperature and has an extrapolated boiling point of 27°. Boron exchange is observed with diborane(6) enriched in boron-10 [109) and the SH proton exchanges with deuterium chloride [109]. The presence of the mercapto-group in the bridging position is unequivocally established by the boron-11 nmr spectrum. A 1:2:1 triplet is observed with each member of the triplet exhibiting well resolved doublet fine structure due to coupling with the bridge hydrogen.

b) μ-Dimethylaminodiborane(6)

A number of Lewis base adducts of μ-dimethylaminodiborane(6) have been prepared [110]. Based upon the observed extent of dissociation of

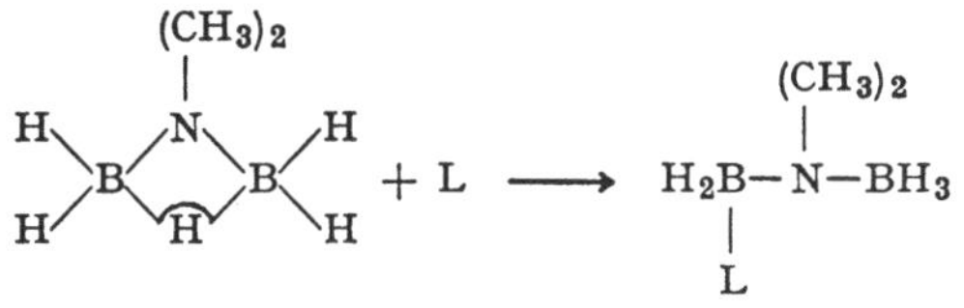

these adducts in the vapor phase, the following relative order of base strengths is reported [110]: $C_5H_5N > (CH_3)_3P > 2\text{-}(CH_3)C_5H_4N > (CH_3)_2PH > CH_3PH_2 > CF_3(CH_3)_2P$. Adducts of the latter three bases are completely dissociated in the gas phase. The reversible decomposition reaction shown below is observed when either CH_3PH_2 or $CF_3(CH_3)_2P$ is the ligand.

$$\mu\text{-}(CH_3)_2NB_2H_5L \rightleftharpoons BH_3L + BH_2N(CH_3)_2$$

It is noted that μ-dimethylaminodiborane(6) is a much weaker Lewis acid than diborane(6).

The rate constant for the intramolecular hydrogen exchange in μ-dimethylaminodiborane(6) has been evaluated by a detailed analysis of the line shape of the boron-11 nmr spectra as a function of temperature [111]. This constant is of the form $K(C,T) = k_1(T) + k_2(C,T)$ where k_1 is the rate constant in inert solvents or the neat liquid, C is the concentration of tetrahydrofuran or 1,2-dimethoxyethane and k_2 is the rate constant for the ether catalyzed exchange. The results are consistent with the mechanism proposed by Gaines and Schaeffer [112] which involves cleavage of the hydrogen bridge followed by rotation of the BH_3 group and reestablishment of the bridge.

c) Alkyl Derivatives of Diborane(6)

The hydroboration of 1,3 butadiene by diborane(6) yields polymeric material [113] which can be thermally cracked [114] to produce a variety of organoboranes and alkyl substituted derivatives of diborane(6) the structures of which have been in dispute. In particular, Köster and

Iwasaki [115] proposed structures containing the borolane ring (a, b, c) for the compounds $(C_4H_6)B_2H_4$, $(C_4H_6)_2B_2H_2$ and $(C_4H_6)_3B_2$ respectively.

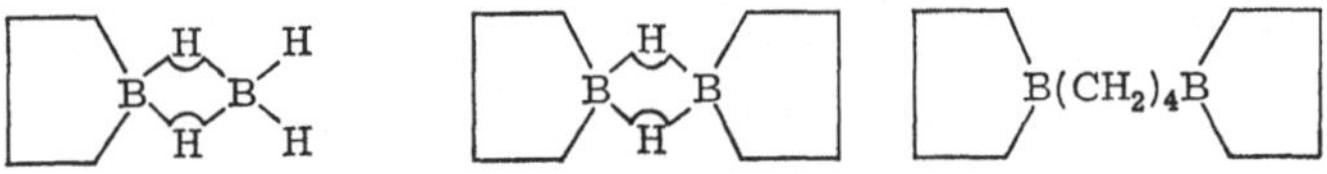

a b c

However, the boron-11 nmr spectrum of $(C_4H_6)B_2H_4$ [116,117] clearly shows that it has structure 1 below, not structure a. Moreover, the chemical properties reported for $(C_4H_6)_2B_2H_2$ [115] are not consistent with those of known tetraalkyl derivatives of diborane(6), *e.g.* at room temperature it does not hydroborate terminal olefins nor react with methanol. Brown and coworkers therefore proposed [114] structure 2 for $(C_4H_6)_2B_2H_2$.

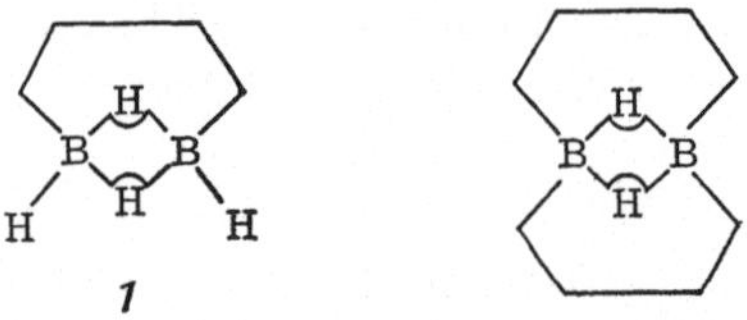

1 2

There is no simple spectroscopic method which will differentiate between structures b and 2, but Young and Shore [12] have demonstrated that structure 2 is indeed the correct formulation for $(C_4H_6)_2B_2H_2$. Consider the symmetrical cleavage products expected from the structures in question:

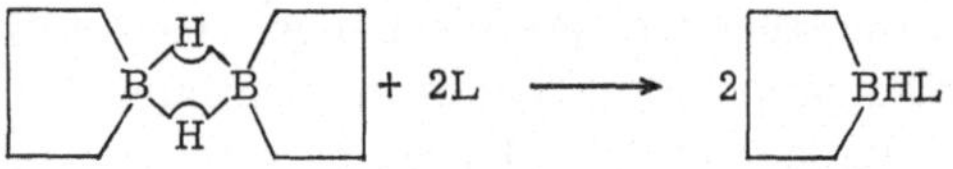

b

$$\tag{1}$$

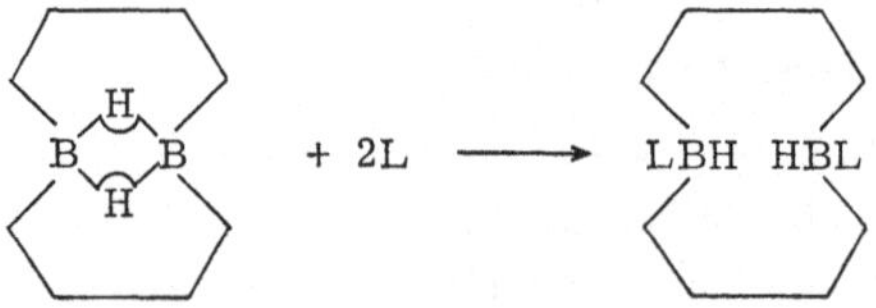

2

$$\tag{2}$$

The molecular weight of the cleavage product of structure *b* (reaction 1) would be one-half as large as that from structure *2* (reaction 2). The molecular weights of $(C_4H_6)_2B_2H_2 \cdot 2\,L$ $(L = (CH_3)_3N, (CH_3)_2NH$ and $(CH_3)NH_2)$ provide convincing evidence for the 1,2-bis(tetramethylene)-diborane(6) formulation (structure *2*). Additional evidence in support of this formulation has also been presented [118].

Brown and co-workers [118] have been unable to find experimental evidence in support of structure *c* for $(C_4H_6)_3B_2$.

Gaines has prepared 1,1-dimethyldiborane(6) in relatively high yield by means of a novel "insertion reaction" which appears to be of general utility.

$$(CH_3)_2BCl + NaBH_4 \longrightarrow 1,1\text{-}(CH_3)_2B_2H_4 + NaCl$$

Long and Wallbridge [120] report the preparation of mixed methyl derivatives of diborane(6) from the reaction of trimethylborane with lithium or sodium tetrahydroborate(−1) and hydrogen chloride in the presence of suitable Lewis acid.

IV. Triborane(7) Adducts

A. Structure

The structure of $B_3H_7 \cdot NH_3$ was determined by Nordman and Reimann [162].

$L = NH_3$

B. Boron-11 NMR

A study of the boron-11 nmr spectra of the tetrahydrofuran and trimethylamine adducts of triborane(7) [121] has shown that the rapid base exchange which renders the boron atoms magnetically equivalent in the first adduct does not occur in the second adduct even though a rapid tautomerization involving all hydrogen atoms is evident in both cases. Thus the spectrum of $B_3H_7 \cdot THF$ in THF or in benzene exhibited one poorly resolved multiplet similiar to that reported by Phillips, Miller and Muetterties for $B_3H_7 \cdot O(C_2H_5)_2$ [122]. The spectrum of $B_3H_7 \cdot N(CH_3)_3$ in ethyl ether or in benzene was consistent with two overlapping octets in the area ratio of 1:2 centered about 35 hz. apart [123].

C. The Preparation and Chemistry of the Dimethyl Ether Adduct of Triborane(7); Diborane(4) Adducts

The dimethyl ether adduct of triborane(7) [124] has been prepared by the cleavage of tetraborane(10) at room temperature according to the equation:

$$B_4H_{10} + (CH_3)_2O \longrightarrow B_3H_7 \cdot O(CH_3)_2 + \tfrac{1}{2} B_2H_6.$$

Removal of diborane(6), unreacted tetraborane(10) and excess dimethyl ether at $-45°$ left the adduct as a white solid (melting point 10.5 to 11.1°).

The adduct decomposed in about an hour at room temperature to a mixture of known boron hydrides. Reaction of the adduct with BF_3 at $-16°$ yielded a somewhat different mixture of boron hydrides. It is of interest that hydrogen gas was not formed in either case.

Cleavage of $B_3H_7 \cdot O(CH_3)_2$ by PF_3 occurs slowly at $-16°$ [125,126]. The major product, $B_2H_4(PF_3)_2$, is very reactive and is potentially a very useful reagent for synthesis. This novel diborane(4) adduct melts at $-114.3°$ and decomposes by a second-order process. It was characterized by vapor density, infra-red spectroscopy and mass spectroscopy. Acid hydrolysis produced about four moles of H_2 and basic hydrolysis produced about five moles of H_2 per mole of $B_2H_4(PF_3)_2$. In addition, the following reactions were found to proceed nearly quantitatively.

$$B_2H_4(PF_3)_2 + H_2 \longrightarrow 2\,BH_3PF_3$$

$$B_2H_4(PF_3)_2 + B_2H_6 \longrightarrow B_4H_{10} + 2\,PF_3$$

The reaction of $B_2H_4(PF_3)_2$ with trimethylamine does not give $B_2H_4[N(CH_3)_3]$ as one would expect; the products are $BH_3N(CH_3)_3$ and $[BHN(CH_3)_3]_n$ [127]. The preparation of $B_2H_4[PF_2N(CH_3)_2]_2$ by the reaction of $PF_2N(CH_3)_2$ with $B_3H_7 \cdot PF_2N(CH_3)_2$ has been accomplished [129] but the details have not yet been published. Other diborane(4) adducts have recently been prepared by new and completely different methods. The coupling of $BH_2Br \cdot N(CH_3)_3$ by sodium-potassium alloy in liquid trimethylamine yields $B_2H_4[N(CH_3)_3]_2$ [127]. This method should be of general utility. The cleavage of $B_5H_9[P(CH_3)_3]_2$ [19] (see discussion under Pentaborane(9)) by excess Lewis base under carefully controlled conditions produces $B_2H_4[P(CH_3)_3]_2$.

D. The Octahydrotriborate(-1) Ion

Transition metal complexes containing the octahydrotriborate(-1) ion have been prepared [130-131]. The x-ray crystal structures of $(OC)_4CrB_3H_8$ [132] and $[(C_6H_5)_3P]_2CuB_3H_8$ [133] reveal that the octahydrotriborate(-1)

ion functions as a bidentate ligand coordinated to the metal by two hydrogen bridges as shown schematically below.

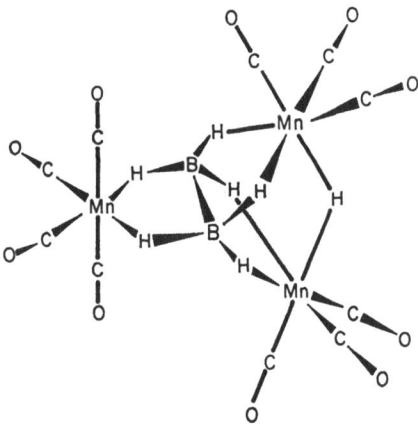

This is the third type of hydrogen bridging to be identified for the metal hydroborates; a double hydrogen bridge from the boron to the metal is found in $[(C_6H_5)_3P]_2CuBH_4$ [134] and triple hydrogen bridges exist in $Zr(BH_4)_4$ [135].

Three sets of double hydrogen bridges are found in $Mn_3(CO)_{10}$ $(BH_3)_2H$. The structure of this unique compound was determined by Kaesz, Fellman, Wilkes, and Dahl [136] and is shown Fig. 2.

Fig. 2. Molecular structure of $Mn_3(CO)_{10}(BH_3)_2H$

Several new ionic octahydrotriborate(−1) salts have been prepared by metathesis reactions [137,138].

The existence of boron-10 spin coupling in the octahydrotriborate(−1) ion has been proved by Norman and Schaeffer [139] through boron-11

121

nmr studies of boron-10 enriched species. As the boron-10 concentration increased, the resolution of the nonet decreased; however, by spin decoupling at the boron-10 frequency, resolution could be restored.

V. Tetraborane(8) Adducts

The question of the molecular structure of the adducts of tetraborane(8) has now been resolved [144]. The boron-11 nmr spectra of the carbon monoxide and trifluorophosphine adducts have been analyzed [140] in terms of the 2112 topological geometry [141] proposed by Dupont and Schaeffer [142]. The dimethylaminodifluorophosphine adduct of tetraborane(8) [143] which was prepared by the following reaction is stable at room temperature.

$$B_4H_8CO + (CH_3)_2NPF_2 \xrightarrow{-20°} B_4H_8 \cdot PF_2N(CH_3)_2 + CO$$

The structure of this adduct has been determined by X-ray diffraction [144] and confirms the 2112 topology of the B_4H_8 framework. Furthermore, in agreement with the original conclusion [143], the adduct contains a boron-phosphorus bond rather than a boron-nitrogen bond.

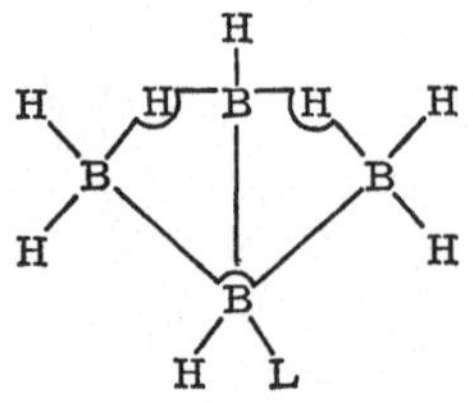

$L = PF_2N(CH_3)_2$

The recently published fluorine-19 nmr spectrum of $B_4H_8 \cdot PF_2$ $(CH_3)_2$ [145] indicates that two isomers are present at room temperature. Rapid interconversion occurs at 90° to 100° as shown by the reversible coalescence of the fluorine-19 resonance signals.

Difluorophosphine displaces carbon monoxide from B_4H_8CO to give the corresponding $B_4H_8PF_2H$ which is stable at room temperature [145]. The fluorine-19, boron-11, phosphorous-31 and proton nmr spectra reveal no isomers, but show that the boron-phosphorus and boron-hydrogen coupling constants are approximately equal in this adduct as is the case for $B_4H_8PF_2N(CH_3)_2$ [145] and $B_4H_8PF_3$ [140].

VI. Tetraborane(10)

A. Preparation

The preferred method for the preparation of B_4H_{10} in the laboratory is the protolysis of $(CH_3)_4NB_3H_8$ by 85 per cent H_3PO_4 reported by Gaines and Scheaffer [168]. The yield is 40 per cent and $(CH_3)_4NB_3H_8$ is readily available [169].

B. The Reaction with NaBD₄ in 1,2-Dimethoxyethane

A tracer study of the reaction of tetraborane(10) with sodium tetra-deuteroborate(-1) in dimethoxyethane (glyme) at $-45°$ has shown that the products of symmetrical cleavage are produced [146]:

$$NaBD_4 + B_4H_{10} \longrightarrow NaB_3H_7D + B_2H_3D_3$$

The authors propose a mechanism involving attack of the tetra-deuteroborate(-1) ion on a single hydrogen-bridged intermediate, glyme$\cdot BH_2-H-B_3H_7$, resulting in the transfer of the deuteride ion causing the displacement of the $BH_3\cdot$glyme moiety. It is pointed out that symmetrical cleavage of tetraborane(10) by the solvent followed by deuteride ion transfer from tetradeuteroborate(-1) to the triborane (7)$\cdot$glyme adduct is also consistent with the experimental results.

C. Deuterated Derivatives

Two specifically deuterated tetraborane(10) derivatives have been prepared. Degradation of pentaborane(11) by D_2O [147,148] gives 95—97% yields of bridge monodeuterated tetraborane(10). Unsymmetrical cleavage is suggested as a possible mechanism:

$$B_5H_{11} + 2D_2O \longrightarrow [BH_2(OD_2)^+_{\frac{1}{2}}B_4H_9^-] \xrightarrow{D_2O} \mu\text{-}DB_4H_9 + B(OD)_3 + 2HD$$

The reaction of the carbon monoxide adduct of tetraborane(8) with deuterium gas gives only dideuterotetraborane(10) which has one bridging deuterium and a terminal deuterium at the 1 position [140]:

$$B_4H_8CO + D_2 \longrightarrow \mu,1\text{-}D_2B_4H_8$$

The determination of the location of the deuterium atoms in these two compounds was accomplished by the use of infra-red [149] and boron-11 nmr spectroscopy. These techniques also showed that complete scrambling occurred in about 1.3 hours at room temperature and the authors note that the previously prepared deuterium substituted tetraborane(10) [150,151] was therefore probably not specifically labeled. This conclusion is consistent with later zero source contact mass spectral studies [152].

D. NMR Spectra

The boron-11 nmr spectrum of tetraborane(10) has been reinterpreted [153,154]. The proposal that the fine structure on the high field doublet is the result of first order boron-10 spin coupling with boron-11 across the B_1—B_3 bond [155] has been refuted. Completely deuterated tetraborane(10) does not exhibit such fine structure [153] and double resonance experiments at the boron-10 resonance frequency on normal tetraborane(10) do not affect the boron-11 nmr spectrum [153]. Boron-10 nmr of normal and boron-10 enriched tetraborane(10), boron-11 nmr of boron-10 enriched tetraborane(10), and boron-11 nmr of normal tetraborane(10) with decoupling at the proton frequency are all consistent with the conclusion that second order effects are responsible for the observed fine structure on the high field doublet. Calculations assuming only proton and boron-11 coupling reproduce the boron-11 and proton nmr spectra reasonably well [154]. The boron-10-boron-11 spin-coupling constant is estimated as 17 hz. and it is suggested that quadrupolar relaxation effectively removes such coupling [154]. Boron-11 nmr of boron-10 enriched tetraborane(10) in conjunction with decoupling at the boron-10 frequency shows, however, that boron-10 spin-coupling is a factor to be considered in the analysis of the boron-11 nmr spectrum of tetraborane(10).

E. Mass Spectra

A careful study of the pyrolysis of tetraborane(10) [158] using an "integral furnace" type mass spectrometer in which the resulting ions have zero source contact has provided good evidence for the previously proposed [142,157] but unobserved borane, B_4H_8. Diborane(6) and higher hydrides containing from five to ten boron atoms were also formed. Neither triborane(7) nor monoborane, BH_3, was seen. The same technique was applied to the mass spectral analysis of certain isotopically substituted tetraboranes [152]. It is shown that pyrolysis of tetraborane(10) occurs in the ion source of the conventional mass spectrometer. Fragmentation of the boron framework occurs by non-random processes but the previous proposal that hydrogen and deuterium atoms are lost from specific positions in the dissociation of $[B_4H_x]^+$-ions [151] is not supported. The negative ion mass spectrum of tetraborane(10) has also been reported [159].

F. Alkyl Derivatives of Tetraborane(10)

Two new methods for the preparation of methyl derivatives of tetraborane(10) have been developed. The first method, which is useful on a synthetic scale, yields the new compound $2,2(CH_3)_2B_4H_8$ in 49 per cent yield from the reaction of NaB_3H_8 with $(CH_3)_2BCl$ [119]. This compound is thermally unstable decomposing at a moderate rate at room temperature even in the gas phase at low pressure. The structure is unambiguously deduced from the boron-11 nmr spectrum which is closely related to the spectrum of the parent compound, B_4H_{10}.

The second method [160] involves the reaction of $B_3H_7 \cdot O(CH_3)_2$ with $1,2\text{-}(CH_3)_2B_2H_4$ and is conducted on a much smaller scale. It yields a mixture of products requiring the skillful application of glc techniques for separation; however, the results are of significant chemical interest. The major product of this reaction is the previously known $2\text{-}(CH_3)B_4H_9$ [161]; minor products include the previously unknown $2,4\text{-}(CH_3)_2B_4H_8$ and $1,2\text{-}(CH_3)_2B_4H_8$ along with traces of a material which is thought to be $2,2\text{-}(CH_3)_2B_4H_8$. The 2,4 and 1,2 isomers were definitely identified by means of boron-11 nmr spectroscopy, but so little of the 2,2 isomer was produced that nmr studies were precluded. The identification of this material as the 2,2 isomer as opposed to the remaining possibility, $1,3\text{-}(CH_3)_2B_4H_8$, is based on a comparison of the mass spectral cracking patterns of the three isomers. The agreement between the infra-red spectrum of $2,2\text{-}(CH_3)B_4H_8$ reported here and that reported [119] for the more fully characterized material prepared by Gaines is not entirely satisfactory, but this could be due to the poor stability of the compound [119].

The formation of $2\text{-}(CH_3)B_4H_9$ is thought to arise from the combination of B_3H_7 with $(CH_3)BH_2$. The dimethyl derivatives are believed to result from the following process [160]:

$$B_3H_7 \cdot O(CH_3)_2 \longrightarrow [B_2H_4] + BH_3 \cdot O(CH_3)_2$$

$$[B_2H_4] + 1,2\text{-}(CH_3)_2B_2H_4 \longrightarrow (CH_3)_2B_4H_8$$

Consistent with this concept is the observation that in the presence of BF_3, the reaction of $B_3H_7 \cdot O(CH_3)_2$ with $1,2\text{-}(CH_3)_2H_4$ produces no $(CH_3)_2B_4H_8$ while the yield of $2\text{-}(CH_3)B_4H_9$ is virtually unchanged. Furthermore, the reaction of $B_2H_4(PF_3)_2$ with $1,2(CH_3)_2B_2H_4$ has been shown to produce mixed $(CH_3)_2B_4H_8$ isomers as the major products with B_4H_{10} and $2\text{-}(CH_3)B_4H_9$ also being formed [126].

Tetraborane(10) reacts with $RC\equiv CR$ (R=H, CH_3) to produce mixtures of small nido-(open cage) or closo-(closed cage) carboranes [163]; however, with C_2H_4, 2,4-dimethylenetetraborane(10) is formed [164]. This compound is properly classified as an alkyl substituted tetraborane(10).

When C_2D_4 is reacted with B_4H_{10}, the product is $D_4C_2B_4H_8$ as shown by infra-red and boron-11 nmr studies [165]. This supports the following mechanism:

$$B_4H_{10} \longrightarrow [B_4H_8] + H_2$$
$$[B_4H_8] + C_2D_4 \longrightarrow D_4C_2B_4H_8$$

G. 2,2'-Bitetraboranyl

Available evidence favors the formulation of B_8H_{18} as 2,2'-bitetra-boranyl [166,167]. This borane is produced in trace amounts by the reaction

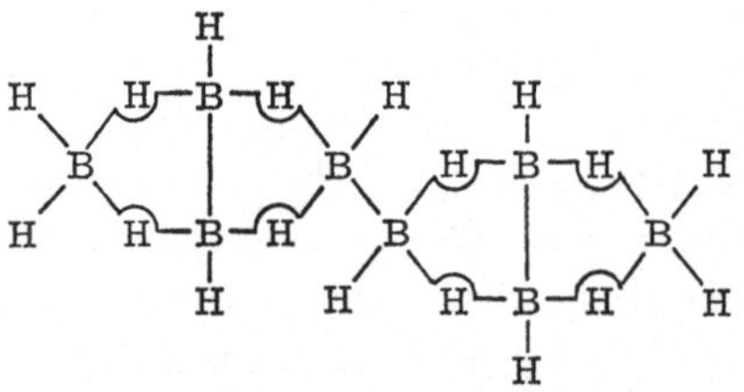

of $(CH_3)_4NB_3H_8$ with polyphosphoric acid *in vacuo* [166]. The boron-11 nmr spectrum [166] was originally interpreted in terms of this structure, but an alternative beltline arrangement of boron atoms could not be definitely excluded on this basis. A very thorough zero source contact mass spectral study [167] has provided strong evidence in favor of the H_9B_4–B_4H_9 arrangement as opposed to the beltline arrangement.

VII. Pentaborane(9)

A. The Octahydropentaborate(-1) Ion

a) Synthesis and Characterization

Although it was reported that pentaborane(9) did not react with sodium hydride in ethyl ether [170], more recently, Onak, Dunks, Searcy, and Spielman [14] found that sodium hydride and lithium hydride do react slowly at room temperature in ethyl ether. More basic ethers such as diglyme (bis(2-methoxyethyl)ether) or tetrahydrofuran increase the rate of reaction; up to 80 per cent of the hydrogen expected from the following reaction was obtained.

$$NaH + B_5H_9 \longrightarrow NaB_5H_8 + H_2$$

The boron-11 nmr spectrum of the product obtained by Onak *et al.* bore no resemblence to that of pentaborane(9) and reaction of the product with anhydrous acid did not yield pentaborane(9).

The conjugate base of pentaborane(9), $B_5H_8^-$, was prepared and characterized independently by Gaines and Iorns [15] and by Geanangel and Shore [16]. Gaines and Iorns used alkyl lithium compounds as deprotonating agents in ethyl ether at low temperature and obtained LiB_5H_8 as a non-volatile strongly solvated, colorless, viscous oil. The reaction of this product with excess anhydrous hydrogen chloride regenerated pentaborane(9) in 72 per cent yield and the same reaction with deuterium chloride produced μ-DB_5H_8. The mass spectrum of this product showed monodeuteration only; furthermore, ^1H nmr and infra-red studies proved that the deuterium atom was present at the bridging position only. In the presence of weak Lewis bases such as tetrahydrofuran and ethyl ether, a slow intramolecular exchange was observed between the bridging deuterium and the basal terminal hydrogens. The boron-11 nmr spectrum of LiB_5H_8 contained a high field doublet of relative intensity 1 similiar in appearance and chemical shift to that observed in the spectrum of pentaborane(9) and it was therefore assigned to an apical boron in the framework of the ion [15]. This fact, in conjunction with the previously discussed results shows that the structure of the anion is similiar [15] to the structure of the parent hydride, B_5H_9. The presence in the boron-11 nmr spectrum of an asymmetric, poorly resolved group of peaks with a relative intensity of four led the authors to propose that more than two types of basal boron atoms were present in solution [15].

Geanangel and Shore [16] obtained the boron-11 nmr spectra of samples of LiB_5H_8, NaB_5H_8, and KB_5H_8 which were prepared and maintained at low temperature. Two symmetrical doublets in the area ratio of four to one were observed for NaB_5H_8 and KB_5H_8 at low temperature, but the compounds were reported to be of poor thermal stability. At room temperature, ether solutions of LiB_5H_8 were stable for about an hour and the boron-11 nmr spectrum was identical with that of KB_5H_8; however, at progressively lower temperatures, the basal doublet decreased in relative intensity [16] and lost resolution [16,17]. The boron-11 nmr spectrum of $(CH_3)_4NB_5H_8$, prepared by metathesis from LiB_5H_8, did not resemble the spectra of the alkali metal salts of $B_5H_8^-$; it was apparently a structural isomer [16].

Further studies reported by Johnson, Geanangel and Shore [17] have led to the isolation of KB_5H_8 as a microcrystalline white powder of limited thermal stability. The temperature dependence of the LiB_5H_8 ^{11}B nmr spectrum (Fig. 3) was ascribed to boron nuclear quadrupolar relaxation effects caused by the increased viscosity of the solution at lower temperatures [17]. The spectrum of the potassium salt showed the same temperature dependence as LiB_5H_8 but at temperatures 60 to 70 degrees lower [17]. The magnetic equivalence of the basal boron atoms

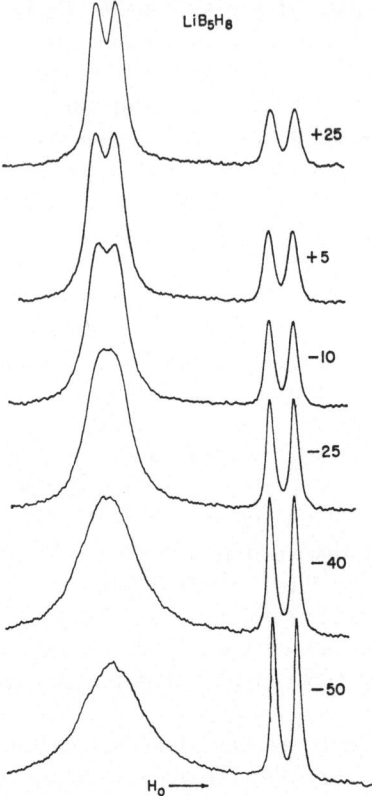

Fig. 3. Boron-11 nmr spectrum of LiB_5H_8 in ethyl ether at 32.1 Mhz

in spite of the presumed structural non-equivalence is undoubtedly due to a rapid hydrogen atom tautomerism in solution. A mechanism of tautomerism like that proposed [171] for the analogous species, hexaborane(10), is believed to be operative [17] (see Fig. 4).

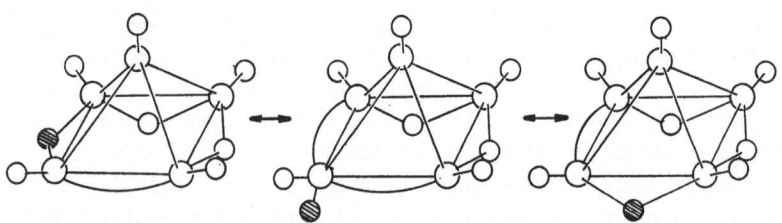

Fig. 4. A mechanism for the tautomerism of the $B_5H_8^-$ ion in solution

The synthesis and characterization of the $[B_5H_8]^-$ ion [15,16,17] is one of the most important developments in recent years in the chemistry of pentaborane(9) and its derivatives. Several examples of a new class of pentaborane(9) derivatives have been prepared in which substituents occupy bridging sites [172–175] and significant yields of hexaborane(10) [16,17], hexaborane(12) [17], or decaborane(14) [16,17] have been obtained from the reaction of $B_5H_8^-$ with B_2H_6 by careful control of the reaction conditions.

b) μ-$R_3MB_5H_8$

The first bridge substituted pentaborane(9) was μ-$(CH_3)_3SiB_5H_8$ prepared by Gaines and Iorns [172] from $(CH_3)_3SiCl$ and LiB_5H_8. Other analogous Group IV derivatives of pentaborane(9) have been prepared by this general reaction [173].

$$LiB_5H_8 + R_3MCl \longrightarrow \mu\text{-}R_3MB_5H_8 + LiCl$$

$$R_3M = H_3Si,\ (CH_3)_3Si,\ (C_2H_5)_3Si,\ H_3Ge,\ (CH_3)_3Ge,$$

$$(C_2H_5)_3Ge,\ (CH_3)_3Sn,\ (CH_3)_3Pb$$

Evidence of the bridging location of the Group IV atom was obtained in each case from boron-11 and proton nmr studies; furthermore, preliminary results of X-ray diffraction studies on 1-Br, μ-$(CH_3)_3SiB_5H_7$ show that the silicon atom does indeed bridge two basal boron atoms [177]. The bridged silicon and germanium derivatives isomerize to the corresponding 2-substituted pentaborane(9) derivatives in the presence of weak Lewis bases at room temperature thereby showing the greater thermodynamic stability of the latter isomer. The formation of the less stable bridged derivatives in the reactions of the $[B_5H_8]^-$ ion with these Group IV halides is thus strong evidence that the site of Lewis basicity in the tetragonal pyramidal structure for the $[B_5H_8]^-$ ion, proposed by Johnson, Geanangel and Shore [16,17], is the bent boron-boron bond in the base.

c) μ-$R_2PB_5H_8$

The preparative method of Gaines and Iorns [172,173] discussed above has been extended by Burg and Heinen [174] in the syntheses of some phosphino-pentaborane(9) compounds. Two isomers of μ-$(CF_3)(CH_3)PB_5H_8$, A and B, were prepared and characterized. Compound A, which was preferentially formed, isomerized to compound B by a first order process.

It is concluded that one isomer has the methyl group below the basal plane of the boron framework (*endo*-) with the CF$_3$ group nearly in the plane of the base in an axial (*exo*-) position and in the other isomer the conformation is the opposite, but no assignment was made. Marynick and Onak [180] have suggested, on the basis of their ring current model for the correlation of chemical shifts in pyramidal boron compounds, that the proton chemical shifts reported for the methyl groups of these two isomers [174] favor assignment of the *endo*-methyl conformation to isomer A.

Also prepared was μ-(CH$_3$)$_2$PB$_5$H$_8$ [174], but the reaction of (CF$_3$)$_2$PI with LiB$_5$H$_8$ yielded only 1-(CF$_3$)$_2$PB$_5$H$_8$. This was attributed to the strongly electron withdrawing character of the bis-trifluoromethyl-phosphino group.

The μ-R$_2$PB$_5$H$_8$ compounds discussed above apparently do not isomerize to the corresponding 2-substituted derivatives as the silicon and germanium derivatives do. This is not unexpected in view of the fact that the bridging phosphino-group is connected to the two basal borons by two 2-center two electron bonds as opposed to the one 3-center two electron bond which presumably binds the Group IV moieties previously discussed.

d) The Reaction of Diborane(6) with the Octahydropentaborate(−1) Ion

The isolation and identification of hexaborane(10) and decaborane(14) from the reaction of diborane(6) with lithium octahydropentaborate(−1) was reported by Geanangel and Shore [16]. Decaborane(14) was obtained in 25 to 30 per cent yields by refluxing 1,2-dimethoxyethane solutions of the reaction products [16,178] while hexaborane(10) was conveniently isolated in 25 to 35 per cent yields when the reaction was conducted in dimethyl ether and the products were distilled from the reaction mixture at low temperature [176−179]. The reaction was initially thought to involve the unsymmetrical cleavage of diborane(6) by the octahydro-pentaborate(−1) ion because the reaction stoichiometry appeared to be 1 B$_2$H$_6$ to 1 B$_5$H$_8^-$ and both LiBH$_4$ and B$_6$H$_{10}$ were found among the reaction products [16].

Further study has shown that the reaction actually proceeds in two steps [176]. The first step is a rapid and irreversible reaction which gives the product of symmetrical cleavage of diborane(6), the undecahydro-hexaborate(−1) ion. The alkali metal salts of this new ion rapidly decom-

$$B_5H_8^- + \tfrac{1}{2}B_2H_6 \xrightarrow{-78°} B_6H_{11}^-$$

pose in ether solvents above $\sim -15°$; complete solvent removal is not possible at low temperatures. The boron-11 nmr spectrum of B$_6$H$_{11}^-$ is consistent with and evidence for a bridged $B_5H_8^-$ structure; thus, the resonance signal of the basal boron atoms is split into two doublets, each of relative intensity 2 and each at lower field than the basal resonance in the $B_5H_8^-$ ion. The doublet at highest field is of relative intensity 1 and easily assignable to the apical boron atom; and the bridging boron atom gives rise to a singlet in the spectrum. Since the bridging boron atom is very probably bonded to at least two terminal hydrogens, the absence of resolvable spin coupling is disconcerting; however, this could be due to hydrogen atom tautomerism such as frequently simplifies the nmr spectra of boron hydrides.

Diborane(6) reacts with LiB_6H_{11} (the second step in the reaction of B_2H_6 with LiB_5H_8) in a 0.5:1.0 mole ratio but the nature of the product which is formed a low temperature has not been elucidated. On warming, the reaction mixture evolves significant amounts of H_2, B_2H_6, B_5H_9, B_6H_{10} and $B_{10}H_{14}$ while $LiBH_4$ and other unidentified products are found in the non-volatile residue. The thermal decomposition of the $B_6H_{11}^-$ ion in the absence of Lewis acids does not produce any volatile boranes.

The treatment of KB_6H_{11} with hydrogen chloride in ether solution [176] produces B_6H_{12} in yields which are low but comparable to the presently preferred method of preparation [168].

b) μ-$(CH_3)_2BB_5H_8$

Iorns and Gaines [175] have extend the scope of the dimethylboryl insertion reaction [119] in the preparation of μ-$(CH_3)_2BB_5H_8$ from $(CH_3)_2BCl$ and LiB_5H_8. The bridging position of the dimethyl boryl group is clearly shown by the boron-11 nmr spectrum and sp^2 hybridization for the bridging boron is inferred from the extremely low field position of dimethyl boryl resonance. It is interesting and quite significant that this compound isomerizes in ether to a dimethyl derivative of hexaborane(10).

B. Halogen Derivatives

a) Preparation

Prior to 1964, the only halogenated pentaborane(9) derivatives which had been reported were 1-IB_5H_8 and 1-BrB_5H_8 [181,182], but the experimental details of their preparations were sparse.

The preparation of 1-IB_5H_8 in high yield has been accomplished by reacting pentaborane(9) with iodine [183,184], iodine monochloride [185] or iodine trichloride [185]. This compound is a thermally stable crystalline solid which decomposes slowly on contact with the atmosphere [183]. The apically substituted compound is apparently much more stable than the base substituted derivative since the position of equilibrium is far to the left in the following reaction which is catalyzed by Lewis bases [184].

$$1\text{-}IB_5H_8 \;\rightleftharpoons\; 2\text{-}IB_5H_8$$

Nevertheless, since the 2-isomer is much more volatile than the 1-isomer, $2\text{-}IB_5H_8$ may be continually removed from this system and thus isolated as an adequately stable liquid in good yield [184].

The aluminum trichoride catalyzed reaction of bromine with pentaborane(9) [182,183,186] yields exclusively $1\text{-}BrB_5H_8$ [186] and the yields are high. In the absence of the catalyst, a small amount of $2\text{-}BrB_5H_8$ is formed [186] and when UV light is used, the ratio of $2\text{-}BrB_5H_3$ to $1\text{-}BrB_5H_8$ rises to 1.2 although the total yield of isomers is reduced [185]. The preferred method for obtaining $2\text{-}BrB_5H_8$ is the base catalyzed rearrangement of $1\text{-}BrB_5H_8$. Although this equilibrium mixture contains approximately equal amounts of each isomer, they may be readily separated by fractional condensation in the vacuum line since the 2-isomer is significantly more volatile. In view of the statistical factor of 4 which favors the basal isomer, the experimentally observed equimolar mixture of isomers at equilibrium indicates that the apically substituted isomer is the one thermodynamically more stable.

Onak and Dunks were the first to prepare a chloropentaborane(9) in quantities large enough to characterize [187]. The product, isolated in low yield from the reaction of $1\text{-}BrB_5H_8$ with $AlCl_3$ at 150°, was shown to be the basal substituted isomer by boron-11 nmr. It is likely that $1\text{-}ClB_5H_8$ was initially formed and then isomerized to $2\text{-}ClB_5H_8$ under these conditions since $1\text{-}BrB_5H_8$ does not thermally isomerize to $2\text{-}BrB_5H_8$ [187]. Such isomerization is, however, well known for alkyl substituted pentaborane(9) derivatives. The high yield preparation of $1\text{-}ClB_5H_8$ was accomplished by reacting chlorine with pentaborane(9) in liquid boron trichloride with aluminum trichloride at the catalyst [185,188]. The same reaction in the absence of the catalyst is thought to proceed by a free radical mechanism and favors $2\text{-}ClB_5H_8$ but yields are low [185]. This is consistent with a subsequent report on the gas phase reaction of pentaborane(9) with chlorine at low pressure [184].

Ethyl ether promotes the isomerisation of $1\text{-}ClB_5H_8$ to $2\text{-}ClB_5H_8$ and the equilibrium mixture contains about 80 per cent of the latter. Therefore, in view of the previously mentioned statistical factor, the two isomers are about equally stable thermodynamically. The $2\text{-}ClB_5H_8$ thus prepared may be purified by conventional methods in the vacuum line. $1\text{-}ClB_5H_8$ has also been prepared by the reaction of $HgCl_2$ with $1\text{-}IB_5H_8$ in a low pressure flow system [184]; with a conversion of 40 per cent, the yield is 55 per cent. Aluminum trichloride catalyzed chorination of $2\text{-}ClB_5H_8$ yields $1,2\text{-}Cl_2B_5H_7$ as expected [188] and the ether catalyzed isomerization of this compound produces an inseparable mixture of $2,3\text{-}Cl_2B_5H_7$ and $2,4\text{-}Cl_2B_5H_7$ [188].

The only fluoropentaborane known is the 2-substituted derivative which was obtained by the reaction of $2\text{-}IB_5H_8$ with antimony trifluoride in a flow system at low pressure [184]. The compound is a liquid which is moderately stable at room temperature. No boron-fluorine spin coupling is evident in the boron-11 nmr spectrum, but a poorly resolved quartet ($J_{PB} = 60$ hz) is observed in the fluorine-19 nmr spectrum. The analogous reaction of $1\text{-}IB_5H_8$ with antimony trichloride gave a product which was too unstable to characterize [184].

b) Chemistry

Gaines has isolated $2\text{-}(CH_3O)B_5H_8$ from the reaction of $1\text{-}IB_5H_8$ with dimethyl ether [189]. The colorless liquid decomposes at or just below room temperature but in the gas phase at low pressure it is adequately stable. A 1:1 adduct is formed with boron trichloride and on heating the adduct to 50 °C, $2\text{-}ClB_5H_8$ may be isolated [189].

Compounds containing transition metal-boron bonds are produced by the reaction of $2\text{-}ClB_5H_8$ with $NaM(CO)_5$ when M = Re and Mn [190]. These compounds are formulated as $2\text{-}[(CO)_5M]B_5H_8$ on the basis of infra-red and nmr studies [190].

C. Adducts

A thermally stable and slightly volatile adduct of pentaborane(9) has been prepared by the reaction of trimethyl phosphine with pentaborane(9) in ether [19]. The 2:1 reaction stoichiometry was established tensiometrically and the product was characterized by analysis, infra-red spectroscopy, and X-ray powder diffraction. In vacuum, $B_5H_9 \cdot 2\,P(CH_3)_2$ is stable indefinitely at room temperature and sublimes at 115° with negligible decomposition. The boron-11 nmr spectrum consists of two broad unresolved peaks in the area ratio 4:1 (with the smaller peak at higher field). The proton nmr spectrum shows non-equivalent phosphine groups and the $B_2H_2[P(CH_3)_3]_2^+$ ion is present in high abundance in the mass spectrometer; therefore, it is concluded that one ligand is bound to the apical boron and the other to a basal boron in a structure geometrically similiar to that of pentaborane(11) [19] (see structure given in Section II.). A 2:1 composition has previously been reported for the adducts of pentaborane(9) with trialkyl amines [191] and trialkyl phosphines [192,193] prepared by the direct combination of reactants, but these were of poor thermal stability and the structures were not established. The key to the preparation of the stable bis-trimethyl phosphine adduct is the use of the proper solvent. The adduct precipitates from ether as it is formed, but when dissolved in a solvent such as toluene, it reacts with a Lewis base, L, in a 1:1 mole ratio forming 1 mole of $L \cdot BH_3$ per mole of $B_5H_9 \cdot 2 \ P(CH_3)_2$ along with variable amounts of $(CH_3)_3$ PBH_3, $B_2H_4[P(CH_3)_3]_2$ and amorphous material. The bis-trimethyl-phosphinediborane(4) adduct is a new compound and has been fully characterized.

D. Framework Rearrangement

The interconversion of apically substituted and basal substituted derivatives in the presence of Lewis bases or at high temperature is a characteristic feature of pentaborane(9) chemistry and the mechanism has been the subject of considerable speculation [191]. Onak, Friedman, Hartsuck and Lipscomb have found that the product of the base catalyzed isomerization of $1,2-(CH_3)_2B_5H_7$ is exclusively $2,3-(CH_3)_2B_5H_7$ [194,195]; no $2,4-(CH_3)_2B_5H_7$ was detected although it is not expected to be of significantly different stability than the 2,3 isomer. A trigonal bipyramidal transition state is proposed in which the steric requirements of the terminal substituents prevent the formation of the 2,4 isomer. Adduct formation is not considered. However, Onak, Dunks, Searcy and Spielman have shown that the rate of isomerization of selected substituted pentaboranes is dependent on the nature of the base used as the catalyst; therefore, they conclude that the base should be included in the transition state. They also discount the existence of $(CH_3)_3NH^+B_5H_7(CH_3)^-$ which was reported [196] to be the product of the reaction of $1-(CH_3)B_5H_8$ with excess trimethyl amine at room temperature. This anion was believed [196] to be the key intermediate in the base catalyzed isomerization of $1-(CH_3)B_5H_8$ to $2-(CH_3)B_5H_8$. Tuck and Onak [197] have found that 1-Cl, $2-(CH_3)B_5H_7$ and $1-(CH_3),2-ClB_5H_7$ both isomerize to $2-Cl,3-(CH_3)B_5H_7$ in the presence of hexamethylenetetramine.

E. NMR

It has been demonstrated that in the boron-11 nmr spectrum of pentaborane(9), quadrupolar relaxation is the dominant contribution to the spin-lattice relaxation times. The spin-lattice relaxation times were determined [198] and on this basis, the line widths at room temperature are expected to be approximately 0.5 hz. and 5 hz. for the apical and basal resonances respectively. The fact that the observed line widths are about 40 hz. is attributed to the presence of complex, unresolved spin-coupling in the molecule [198].

The proton nmr spectra of most of the boron hydrides are so complex that information is therefore very difficult to extract; however, simplification of the spectrum of $1,2-(CH_3)_2B_5H_7$ by boron-11 multiple spin-decoupling has allowed the complete assignment of its proton nmr spectrum [199]. Three oscillators set at somewhat overlapping frequencies were required to completely spin-decouple all boron atoms. By observing the effect on the proton nmr spectra produced by individual oscillators as well as combinations of oscillators, the assignments could be made unambiguously.

A ring current model has been shown to correlate proton and boron-11 chemical shifts in a number of pyramidal boron compounds including pentaborane(9) and its derivatives [180]. This model is analogous to the model used to explain proton chemical shifts in benzene and similar aromatic compounds.

F. Decaborane(16)

The structure of decaborane(16) [200] reveals that it should be considered as an apically substituted derivative of pentaborane(9). The bond energy of the unique boron-boron bond which connects the two B_5H_8 cages has been determined to be 3.2 ± 0.2 e.v. from mass spectrometry [201] the standard heat of formation is 34.8 kcal. per mole [202]. An analysis of the vibrational spectrum of decaborane(16) has been made by Pinson and Lin [203] who note the absence of the strong absorption at 901 cm^{-1} which was previously reported [204].

VIII. Pentaborane(11)

A. NMR

The boron-11 nmr spectrum of very pure pentaborane(11) taken at 64.2 Mhz. has been obtained and the complete resolution of the low field multiplet into a triplet and a doublet of equal intensity [205] fully confirms the previous analysis of this multiplet based on a spectrum recorded at much lower field strength [206]. The triplet at lowest field is assigned to boron atoms 2 and 5, the doublet at slightly higher field is assigned to the remaining two basal boron atoms, 3 and 4, and the remaining doublet which is at the highest field position is due to the apical boron atom, 1. The 60 Mhz. proton nmr spectrum is also present and assigned [205]. Williams, Gerhart and Pier [205] believe that the unique hydrogen which appears to be in a second apical terminal position from the X-ray structure determination [207] is actually of a more common nature. They state [205] that it is a bridging hydrogen involved in a tautomerism between the equivalent μ-1,2 and μ-1,5 positions (see Fig. 1).

B. Isotopically Labeled Pentaborane(11)

The reaction of B_4H_8CO (enriched in either boron-10 or boron-11) with diborane(6) (enriched in the other isotope) produces pentaborane(11)

135

which is isotopically enriched at basal(2—5) positions [140]. Boron-11 nmr established the position of the labeled boron and showed that completed randomization occurred at room temperature in ten minutes.

IX. Hexaborane(10)

A. Preparation

Study of the chemistry of hexaborane(10) has been severely hampered because of difficulty in preparing this hydride in adequate amounts; however, a recently developed synthesis [179] which provides yields of pure hexaborane(10) of 25 to 35 per cent from commercially available starting materials (see Section VII. A. d.) presages the rapid expansion of research on this compound.

B. Brønsted Acidity

Hexaborane(10) behaves as a Brønsted acid analogous to pentaborane(9) [14,15,16] (see Section VII. A.a.) and decaborane(14) [208]. Strong bases react quantitatively with hexaborane(10) to form the nonahydrohexaborate(−1) ion which is moderately stable at room temperature in an inert atmosphere [18,179].

$$B_6H_{10} + MB \longrightarrow MB_6H_9 + HB$$

$MB = LiCH_3, NaH, KH$

The proton which is removed is a bridge proton as shown by experiments [18] using selectively deuterated hexaborane(10) [209] and the structure thus derived is consistent with the boron-11 nmr spectrum of the nonahydrohexaborate(−1) ion [18,179] which exhibits two symmetrical doublets in the area ratio of five to one (with the smaller peak at higher field). Furthermore, the reaction of the ion with anhydrous hydrogen chloride generates hexaborane(10) in nearly quantitative yield [18,179].

In 1958, Parry and Edwards predicted the acidic character of the bridge hydrogens in the lower boron hydrides and proposed that the acid strength should vary directly with the size of the boron framework for a given analogous series [1]. This has been confirmed for the series B_5H_9, B_6H_{10}, $B_{10}H_{14}$ by proton competition reactions [18] which proceed to

$$B_5H_8^- + B_6H_{10} \longrightarrow B_5H_9 + B_6H_9^-$$

$$B_6H_9^- + B_{10}H_{14} \longrightarrow B_6H_{10} + B_{10}H_{13}^-$$

completion as written as determined by boron-11 nmr. Since decaborane(14) is estimated [210] to be an approximately ten times weaker acid than phenol, the following sequence of Brønsted basicities is indicated:

$$H^- \text{ and } (CH_3)^- > (B_5H_8)^- > (B_6H_9)^- > (B_{10}H_{13})^- > (C_6H_5O)^-.$$

C. Lewis Acidity

Although hexaborane(10) has been reported to form an air stable 1:1 adduct with triphenyl phosphine and a white product with dimethyl sulfide, these products were not characterized [211]. With trimethyl phosphine, however, hexaborane(10) forms a crystalline adduct, $B_6H_{10} \cdot 2 P(CH_3)_3$, the composition of which has been established both tensio-metrically and analytically [20]. The boron-11 nmr spectrum of this thermally stable adduct contains three resonances of equal intensity, a triplet at highest field, a doublet at lowest field, and a broad singlet in between. The two ligands appear magnetically equivalent in the proton nmr spectrum. The adduct is volatile enough to be introduced into the mass spectrometer and a parent ion corresponding to $(B_6H_{10} \cdot 2 P(CH_3)_3]^+$ is seen.

D. Hexaborane(10) as a Base

At low temperatures in a variety of solvents, hexaborane(10) undergoes exchange of a maximum of five hydrogens with deuterodiborane(6) [209]. The product is shown by boron-11 and proton nmr to be substituted only at basal terminal sites. There is no precedent for an exchange of this type at low temperature for a boron hydride without BH_2 groups and it has been proposed [209] that the unique short boron-boron bond in the base of the pentagonal pyramidal framework of hexaborane(10) is involved in the reaction intermediate.

In view of the fact that the boron-boron bond in the base of the structurally homologous octahydropentaborate(-1) ion is a strongly basic site [172,176], it is not unreasonable to suggest that the analogous site in hexaborane(10) has basic properties. It is consistent with this premise that at room temperature, hexaborane(10) exchanges bridge hydrogens rapidly with deuterium chloride *in the gas phase* [176]. Exchange of *terminal* hydrogens occurs between B_5H_9 [212] or $B_{10}H_{14}$ [213] and deuterium chloride and a catalyst is necessary. Clearly, the mechanisms are not the same for these two types of exchange. It is proposed [176] that the B_6H_{10}-DCl exchange proceeds by an associative process which

$$B_6H_{10} + DCl \longrightarrow [B_6H_{10}D]^+Cl^- \longrightarrow \mu\text{-}DB_6H_9 + HCl$$

involves the addition of D^+ to the boron-boron bond in the base of the hexaborane(10) framework. Stability for the $(B_6H_{11})^+$ species has been predicted by Lipscomb on the basis of its structurally and electronically analogous relationship to pentaborane(9) [214].

X. Hexaborane(12)

A. Preparation and Characterization

Quantities of hexaborane(12) large enough to characterize have been isolated by Gaines and Schaeffer [168,215] from the reaction of $(CH_3)_4$ NB_3H_8 with polyphosphoric acid and by Lutz, Phillips and Ritter [216] through the use of gas chromatography on the mixtures obtained from several boron hydride interconversions. The yields were less than five per cent in all cases. The stability of hexaborane(12) is highly dependent on its purity since the material obtained by Gaines and Shaeffer (m.p. —83°, 17 mm Hg at 0°) is reported to decomposed rapidly in the liquid phase while Lutz, Phillips and Ritter found that after three hours at room temperature in the liquid phase, hexaborane(12) could be recovered in 95% yield from samples of the highly purified material (m.p. —82.2, 17.1 mm Hg at 0°) they had originally isolated.

The boron-11 nmr spectrum consists of three resonances of equal intensity. At highest field is a doublet (B_2,B_5) and at lowest field is a doublet (B_3,B_6) with a triplet in between (B_1,B_4) [163] (see topological representation in Fig. 2). The standard heat of formation of hexaborane(12) has been experimentally determined to be 26.5 kcal. per mole[217].

B. Chemistry

Hexaborane(12) reacts with water at 0° to produce tetraborane(10) in over 90 per cent yield and is cleaved by $(CH_3)_2O$ to pentaborane(9) and diborane(6). The latter reaction is considered to be a base catalyzed elimination of a BH_3 group to give a B_5H_9 fragment which rearranges to the known pentaborane(9) [168].

XI. Heptaborane

No heptaborane has been characterized.

XII. Octaborane(12)

A. Preparation

Octaborane(12) was first isolated from a complex mixture produced by a discharge reaction involving pentaborane(9), diborane(6) and hydrogen [218]. Recently, Dobson and Schaeffer have developed a synthesis by which gram quantities of octoborane(12) may be prepared and conveniently isolated [219]. This method involves treatment of KB_9H_{14} [220] with liquid hydrogen chloride to produce the previously unknown iso-B_9H_{15} which decomposes with elimination of hydrogen above —35° to form B_8H_{12}, $B_{10}H_{14}$, $B_{18}H_{22}$ and polymeric material.

B. Chemistry

Diborane(6) enriched in boron-10 isotope reacts with octaborane(12) to from a specifically labeled nonaborane(15) [221]. This is consistent with the proposal of Ditter, Spielman and Williams [222] that octaborane(12), nonaborane(15) and diborane(6) are in equilibrium.

$$B_8H_{12} + \tfrac{1}{2}B_2H_6 \; \rightleftharpoons \; B_9H_{15}$$

They find [222] that in the mass spectrometer, octaborane(12) is produced by the first order decomposition of nonaborane(15) and in turn decomposes by a first order process to hexaborane(10).

Octaborane(12) is a strong Lewis acid forming 1:1 adducts with ethyl ether, trimethyl amine and acetonitrile [223]. With sodium hydride in ether, approximately two thirds of a mole of hydrogen per mole of octaborane(12) is produced but the product is said to be apparently identical with the product of the reaction of sodium amalgam with octaborane(12) which produces no hydrogen [223].

The treatment of the product of the sodium hydride or sodium amalgam reaction described above with liquid hydrogen chloride at low temperature produces an extremely unstable boron hydride which is shown to be octaborane(14) by elemental analysis and boron-11 nmr [223]. This new hydride decomposes quantitatively to octaborane(12) and hydrogen in minutes at room temperature. The structure proposed [223] on the basis of the boron-11 nmr spectrum violates the topological principles based on hydrogen atom crowding [141] and is thus consistent with this facile decomposition.

XIII. References

[1] Parry, R. W., Edwards, L. J.: J. Am. Chem. Soc. 81, 3554 (1959).
[2] Shore, S. G., Hickam, C. W., Jr., Cowles, D.: J. Am. Chem. Soc. 87, 2755 (1965). References to the earlier work of Parry, Schultz, and Shore which established the first example of unsymmetrical cleavage, $BH_2(NH_3)_2^+ BH_4^-$ are given here.
[3] McAchran, G. E., Shore, S. G.: Inorg. Chem. 4, 125 (1965).
[4] Beachley, O. T.: Inorg. Chem. 4, 1823 (1965).
[5] Moews, P. C., Jr., Parry, R. W.: Inorg. Chem. 5, 1552 (1966).
[6] Inoue, M., Kodama, G.: Inorg. Chem. 7, 430 (1968).
[7] Kodama, G., Parry, R. W.: J. Am. Chem. Soc. 82, 6250 (1960).
[8] Hough, W. V., Edwards, L. J.: Advan. Chem. Ser. 32, 189 (1961).

9) Parry, R. W., Rudolph, R. W., Shriver, D. F.: Inorg. Chem. *3*, 1479 (1964).
10) Schaeffer, R., Tebbe, F., Phillips, C.: Inorg. Chem. *3*, 1475 (1964).
11) Edwards, L. J., Hough, W. V., Ford, M. D.: Congr. Intern. Chim. Pure Appl. *1957*, 475 (1958).
12) Young, D. E., Shore, S. G.: J. Am. Chem. Soc. *91*, 3497 (1969).
13) Brown, H. C., Wallace, W. J.: Abstracts, 142nd National Meeting of the American Chemical Society, Atlantic City, New Jersey, 1962, No. 9N. — Wallace, W. J.: Dissertation Abstr. *22*, 425 (1961).
14) Onak, T., Dunks, G. B., Searcy, I. W., Spielman, J.: Inorg. Chem. *6*, 1465 (1967).
15) Gaines, D. F., Irons, T. V.: J. Am. Chem. Soc. *89*, 3375 (1967).
16) Geanangel, R. A., Shore, S. G.: J. Am. Chem. Soc. *89*, 6771 (1967).
17) Johnson II, H. D., Geanangel, R. A., Shore, S. G.: Inorg. Chem. *9*, 908 (1970).
18) — Shore, S. G., Mock, N. L., Carter, J. C.: J. Am. Chem. Soc. *91*, 2131 (1969).
19) Denniston, M., Shore, S. G.: Abstracts, 158th National Meeting of the American Chemical Society, New York, N.Y., Inorganic Division Paper No. 104.
20) — — Unpublished Results.
21) Parry, R. W., Walter, M. K.: Prep. Inorg. React. *5*, 45 (1968).
22) Freeguard, G. F., Long, L. M.: Chem. Ind. (London) *11*, 471 (1965).
23) McCoy, R. E., Bauer, S. H.: J. Am. Chem. Soc. *78*, 2061 (1956).
24) Fehlner, T. P., Koski, W. S.: J. Am. Chem. Soc. *86*, 2733 (1964); *87*, 409 (1965).
25) Baylis, A. B., Pressley, G. A., Jr., Stafford, F. E.: J. Am. Chem. Soc. *88*, 2428 (1966).
26) Wilson, J. H., McGee, H. A., Jr.: J. Chem. Phys. *46*, 1444 (1967).
27) Bauer, S. H.: J. Am. Chem. Soc. *78*, 5775 (1956).
28) Garabedian, M. E., Benson, S. W.: J. Am. Chem. Soc. *86*, 176 (1964).
29) Fehlner, T. P., Koski, W. S.: J. Am. Chem. Soc. *87*, 409 (1965).
30) Grotewald, J., Lissi, E. A., Villa, A. E.: J. Chem. Soc. 1038 (1966).
31) Burg, A. B., Fu, Y. C.: J. Am. Chem. Soc. *88*, 1147 (1966).
32) Fehlner, T. P., Mappes, G. W.: J. Phys. Chem. *73*, 873 (1969).
33) Ganguli, P. S., McGee, H. A., Jr.: J. Chem. Phys. *50*, 4658 (1969).
34) Shore, S. G., Hall, C. L.: J. Am. Chem. Soc. *89*, 3947 (1967).
35) Parry, R. W., Shore, S. G.: J. Am. Chem. Soc. *80*, 15 (1958).
36) Egan, B. Z., Shore, S. G.: J. Am. Chem. Soc. *83*, 4717 (1961).
37) Shore, S. G., Hall, C. L.: J. Am. Chem. Soc. *88*, 5346 (1966).
38) Eastham, J. F.: J. Am. Chem. Soc. *89*, 2237 (1967).
39) Brown, H. C., Tierney, P. A.: J. Am. Chem. Soc. *80*, 1552 (1958).
40) Boker, E. B., Ellis, R. B., Wilcox, W. S.: J. Inorg. Nucl. Chem. *23*, 41 (1961).
41) Gaines, D. F.: Inorg. Chem. *2*, 523 (1963).
42) Phillips, D. A.: Dissertation Abstr. B. *27*, 740 (1966).
43) Matsui, Y., Taylor, R. C.: J. Am. Chem. Soc. *90*, 1363 (1968).
44) Evans, W. G., Holloway, C. E., Surkumarabandhu, K., McDaniel, D. H.: Inorg. Chem. *7*, 1746 (1968).
45) Hall, C. L., Shore, S. G.: Unpublished Results.
46) Malone, L. J., Parry, R. W.: Inorg. Chem. *6*, 817 (1967).
47) Carter, J. C., Parry, R. W.: J. Am. Chem. Soc. *87*, 2354 (1965).
48) Parry, R. W., Nordman, C. E., Carter, J. C., Ter Haar, G.: Advan. Chem. Ser. *42*, 302 (1964).
49) Hellstrom, M., Akerfeldt, S.: Acta. Chem. Scand. *20*, 1418 (1966). — Akerfeldt, S., Wahlberg, K., Hellstrom, M.: Acta Chem. Scand. *23*, 115 (1969). — Williams, R. L.: Acta. Chem. Scand. *23*, 149 (1969).

50) Aylett, B. J., Peterson, L. K.: J. Chem. Soc. 4043 (1965).
51) Gatti, A. R., Wartik, T.: Inorg. Chem. 5, 329 (1966).
52) — — Inorg. Chem. 5, 2075 (1966).
53) Belinski, C., Francaes, G., Horney, C., Lalau-Keraly, F.: Compt. Rend. 259, 3737 (1964).
54) Pattison, I., Wade, K.: J. Chem. Soc. A 842 (1968).
55) Miller, N. E.: J. Am. Chem. Soc. 88, 4284 (1966).
56) Craig, R. H., Greenwood, N. N.: J. Chem. Soc. A 961 (1967).
57) Walker, F. E., Pearson, R. K.: J. Inorg. Nucl. Chem. 27, 1981 (1965).
58) Rudolph, R. W., Parry, R. W.: J. Am. Chem. Soc. 89, 1621 (1967).
59) Kodama, G., Parry, R. W.: Inorg. Chem. 4, 410 (1965).
60) Morse, K. W., Parry, R. W.: J. Am. Chem. Soc. 89, 172 (1967).
61) Laube, B. L., Bertrand, R. D., Casedy, G. A., Compton, R. D., Verkade, J. G.: Inorg. Chem. 6, 173 (1967).
62) Nöth, H., Schraegle, W.: Chem. Ber. 98, 352 (1965).
63) Drake, J. E., Riddle, C.: J. Chem. Soc. A 1675 (1968).
64) — Simpson, J.: Inorg. Chem. 6, 1984 (1967).
65) — Goddard, N.: J. Chem. Soc. A 662 (1969).
66) Laurent, J. P., Jugie, G., Comenges, G.: J. Inorg. Nucl. Chem. 31, 1353 (1969).
67) — Jugie, G., Wolfe, R., Commenges, G.: J. Chim. Phys. Physicochim. Biol. 66, 409 (1969).
68) Gilje, J. W., Morse, K. W., Parry, R. W.: Inorg. Chem. 6, 1761 (1967).
69) Thompson, N. R.: J. Chem. Soc. 6290 (1965).
70) Mayer, E., Hester, R. E.: Spectrochim. Acta A 25, 237 (1969).
71) Hester, R. E., Mayer, E.: Spectrochim. Acta A 23, 2218 (1967).
72) Payne, D. S., Noeth, H., Henniger, G.: Chem. Commun. 327 (1965).
73) Lundburg, K. L., Rowatt, R. J., Miller, N. E.: Inorg. Chem. 8, 1336 (1969).
74) Wagner, R. I., Wilson, C. O., Jr.: Inorg. Chem. 5, 1009 (1966).
75) Riess, J. G., VanWazer, J. R.: J. Am. Chem. Soc. 89, 851 (1967).
76) Kodama, G., Kondo, H.: J. Am. Chem. Soc. 88, 2045 (1966).
77) Riess, J., VanWazer, J. R.: Bull. Soc. Chim. France 1846 (1966).
78) Cowley, A. H., Dierdorf, D. S.: J. Am. Chem. Soc. 91, 6609 (1969).
79) Rudolph, R. W., Parry, R. W., Farran, C. F.: Inorg. Chem. 5, 723 (1966).
80) McGandy, E. L.: Dissertation Abstr. 22, 754 (1961).
81) Sawodny, W., Goubeau, J.: Z. Anorg. Allgem. Chem. 356, 289 (1968).
82) Gamble, E. L., Gilmont, P.: J. Am. Chem. Soc. 62, 717 (1940).
83) Mullikan, R. S.: J. Chem. Phys. 3, 635 (1935).
84) Rustad, D. S., Jolly, W. L.: Inorg. Chem. 7, 213 (1968).
85) Parshall, G. W.: J. Am. Chem. Soc. 86, 361 (1964).
86) Kaesz, H. D., Fellmann, W., Wilkes, G. R., Dahl, L. F.: J. Am. Chem. Soc. 87, 2753 (1965).
87) Shriver, D. F.: J. Am. Chem. Soc. 85, 1405 (1963).
88) Douglass, J. E.: J. Am. Chem. Soc. 86, 5431 (1964).
89) Muetterties, E. L.: Pure Appl. Chem. 10, 53 (1965).
90) Miller, N. E., Muetterties, E. L.: J. Am. Chem. Soc. 86, 1033 (1964).
91) Nainan, K. C., Ryschkewitsch, G. E.: Inorg. Chem. 7, 1316 (1968).
92) Nöth, H., Beyer, H., Vetter, H. J.: Chem. Ber. 97, 110 (1964).
93) Makosky, C. W., Galloway, G. L., Ryschkewitsch, G. E.: Inorg. Chem. 6, 1972 (1967).
94) Ryschkewitsch, G. E.: J. Am. Chem. Soc. 89, 3145 (1967).
95) Jolly, W. L., Schmitt, T.: J. Am. Chem. Soc. 88, 4282 (1966).
96) — — Inorg. Chem. 6, 344 (1967).

97) Miller, N. E., Chamberland, B. L., Muetterties, E. L.: Inorg. Chem. *3*, 1064 (1964).
98) Nöth, H., Beyer, H.: Chem. Ber. *93*, 2251 (1960).
99) Nainan, K. C., Ryschkewitsch, G. E.: J. Am. Chem. Soc. *91*, 330 (1969).
100) Miller, N. E., Resnicek, D. L., Rowatt, R. J., Lundberg, K. R.: Inorg. Chem. *8*, 862 (1969).
101) — Inorg. Chem. *8*, 1693 (1969).
102) Smith, G. L., Kelly, H. C.: Inorg. Chem. *8*, 2000 (1969).
103) Ryschkewitsch, G. E.: J. Am. Chem. Soc. *91*, 6535 (1969).
104) — Garrett, J. M.: J. Am. Chem. Soc. *90*, 7234 (1968).
105) Rowatt, R. J., Miller, N. E.: J. Am. Chem. Soc. *89*, 5509 (1967).
106) Cotton, J. D., Waddington, T. C.: J. Chem. Soc. A, 789 (1966).
107) Keller, P. C.: Chem. Commun. 209 (1969).
108) — Inorg. Chem. *8*, 1695 (1969).
109) — Inorg. Chem. *8*, 2457 (1969).
110) Burg, A. B., Sandhu, J. S.: Inorg. Chem. *4*, 1467 (1965).
111) Schirmer, R. E., Noggle, J. H., Gaines, D. F.: J. Am. Chem. Soc. *91*, 6240 (1969).
112) Gaines, D. F., Schaeffer, R.: J. Am. Chem. Soc. *86*, 1505 (1964).
113) Zweifel, G., Nagase, K., Brown, H. C.: J. Am. Chem. Soc. *84*, 183 (1962).
114) Brown, H. C.: Hydroboration, pp. 209—212. New York: W. A. Benjamin, Inc. 1962.
115) Köster, R.: Angew. Chem. *71*, 520 (1959); *72*, 626 (1960). — Köster, R.: Iwasaki, K.: Boron-Nitrogen Chemistry, pp. 148—165. Advan. Chem. Ser. No. 42. Washington: American Chemical Society 1964. — Köster, R.: Progress in Boron Chemistry, Vol. 1, Chapter 7. H. Steinburg and A. L. McClosky, Ed. New York: Macmillan Co. 1964. — Köster, R.: Advan. Organometal. Chem. *2*, 257 (1964).
116) Weiss, H. G., Lehmann, W. J., Shapiro, I.: J. Am. Chem. Soc. *84*, 3840 (1962).
117) Lindner, H. H., Onak, T.: J. Am. Chem. Soc. *88*, 1886 (1966).
118) Brewer, E., Brown, H. C.: J. Am. Chem. Soc. *91*, 4164 (1969).
119) Gaines, D. F.: J. Am. Chem. Soc. *91*, 6503 (1969).
120) Long, L. J., Wallbridge, M. G. H.: J. Chem. Soc. A 3513 (1965).
121) Ring, M. A., Witucki, E. F., Greenough, R. C.: Inorg. Chem. *6*, 395 (1967).
122) Phillips, W. D., Miller, H. C., Muetterties, E. L.: J. Am. Chem. Soc. *81*, 4496 (1959).
123) The chemical shifts which are reported in reference 121 appear to be in parts per million instead of the indicated cycles per second.
124) Deever, W. R., Ritter, D. M.: Inorg. Chem. *7*, 1036, (1968).
125) — — J. Am. Chem. Soc. *89*, 5073 (1967).
126) — Lory, E. R., Ritter, D. M.: Inorg. Chem. *8*, 5073 (1967).
127) — Miller, F. M., Lory, E. R., Miller, D. M.: Abstracts, The 155th Meeting of the Am. Chem. Soc., San Francisco, April, 1968, M 215.
128) Fleming, M. A.: Dissertation Abstr. *24*, 1385 (1963).
129) Footnote 15, reference 6.
130) Klanberg, F., Muetterties, E. L., Guggenberger, L. J.: Inorg. Chem. *7*, 2272 (1968) and references therein.
131) Lippard, S. J., Ucko, D.: Inorg. Chem. *7*, 1051 (1968) and references therein.
132) Klanberg, F., Guggenberger, L. J.: Chem. Commun. 1293 (1967).
133) Lippard, S. J., Melmed, K. M.: Inorg. Chem. *8*, 2755 (1969).
134) — — Inorg. Chem. *6*, 2223 (1967); J. Am. Chem. Soc. *89*, 3929 (1967).
135) Bird, P. H., Churchhill, M. R.: Chem. Commun. 403 (1967).

[136] Kaesz, H. D., Fellmann, W., Wilkes, G. R., Dahl, L. F.: J. Am. Chem. Soc. *87*, 2753 (1965).

[137] Amburger, E., Gut, E.: Chem. Ber. *101*, 1200 (1968).

[138] Walker, F. E., Pearson, R. K.: J. Inorg. Nucl. Chem. *27*, 1981 (1965).

[139] Norman, A. D., Schaeffer, R.: J. Phys. Chem. *70*, 1662 (1966).

[140] — — J. Am. Chem. Soc. *88*, 1143 (1966); this reaction was first carried out with materials of normal isotopic abundance by Spielman, J. R., Burg, A. B.: Inorg. Chem. *2*, 1139 (1963).

[141] Lipscomb, W. N.: Boron Hydrides. New York: W. A. Benjamin 1963.

[142] Dupont, J. A., Schaeffer, R.: J. Inorg. Nucl. Chem. *15*, 310 (1960).

[143] Ter Haar, G., Fleming, M. A., Parry, R. W.: J. Am. Chem. Soc. *84*, 1767 (1962).

[144] LaPrade, M. D., Nordman, C. E.: Inorg. Chem. *8*, 1669 (1969).

[145] Centofanti, L. F., Kodama, G., Parry, R. W.: Inorg. Chem. *8*, 2072 (1969).

[146] Parry, R. W., Rudolph, R. W., Shriver, D. W.: Inorg. Chem. *3*, 1479 (1964).

[147] The reaction of B_5H_{11} with H_2O was reported by J. L. Boone and A. B. Burg. J. Am. Chem. Soc. *80*, 1519 (1958).

[148] Norman, A. D., Schaeffer, R.: Inorg. Chem. *4*, 1225 (1965).

[149] For an analysis of the vibrational spectrum of B_4H_{10} and B_5H_{11}, see A. J. Dahl, Ph. D. Thesis, University of Michigan, 1963.

[150] Todd, J. E., Koski, W. S.: J. Am. Chem. Soc. *81*, 2319 (1959).

[151] Fehlner, T. P., Koski, W. S.: J. Am. Chem. Soc. *85*, 1905 (1963).

[152] Norman, A. D., Schaeffer, R., Baylis, A. B., Pressley, G. A., Jr., Stafford, F. E.: J. Am. Chem. Soc. *88*, 2151 (1966).

[153] Rigden, J. S., Hopkins, R. C., Baldeschwieler, J. D.: J. Chem. Phys. *35*, 1532 (1961).

[154] Hopkins, R. C., Baldeschwieler, J. D., Schaeffer, R., Tebbe, F. N., Norman, A. D.: J. Chem. Phys. *43*, 975 (1965).

[155] Williams, R. E., Gibbins, S. G., Shapiro, I.: J. Am. Chem. Soc. *81*, 6164 (1959).

[156] Norman, A. D., Schaeffer, R.: J. Phys. Chem. *70*, 1662 (1966).

[157] Brennan, G. L., Schaeffer, R.: J. Inorg. Nucl. Chem. *20*, 205 (1961).

[158] Baylis, A. B., Pressley, G. A., Jr., Gordon, M. E., Stafford, F. E.: J. Am. Chem. Soc. *88*, 929 (1966).

[159] Munro, F., Ahnell, J. E., Koski, W. S.: J. Phys. Chem. *72*, 2682 (1968).

[160] Deever, W. R., Ritter, D. M.: Inorg. Chem. *8*, 2461 (1969).

[161] Lutz, C. A., Ritter, D. M.: Can. J. Chem. *41*, 1344 (1963).

[162] Nordmann, C. E., Reimann, C.: J. Am. Chem. Soc. *81*, 3558 (1959).

[163] Grimes, R. N., Bramlett, C. L., Vance, R. L.: Inorg. Chem. *8*, 55 (1969); *7*, 1066 (1968) and references therein.

[164] Harrison, B. C., Solomon, I. J., Hites, R. D., Klein, M. J.: J. Inorg. Nucl. Chem. *14*, 195 (1960). — Gibbons, S. G., Shapiro, I., Williams, R. E.: J. Phys. Chem. *65*, 1061 (1961).

[165] Williams, R. E., Gerhardt, F. J.: J. Organometal. Chem. *10*, 168 (1967).

[166] Dobson, J., Gaines, D., Schaeffer, R.: J. Am. Chem. Soc. *87*, 4072 (1965).

[167] Steck, S. J., Pressley, G. A., Jr., Stafford, F. E., Dobson, J., Schaeffer, R.: Inorg. Chem. *8*, 830 (1969).

[168] Gaines, D. F., Schaeffer, R.: Inorg. Chem. *3*, 438 (1964).

[169] — — Tebbe, F.: Inorg. Chem. *2*, 526 (1963).

[170] Hough, W. V., Edwards, L. J.: Advan. Chem. Ser. *32*, 191 (1961). — Furukawa, G. T., Park, R. P.: J. Res. Natl. Bur. Std. *55*, 255 (1955).

[171] Reference 141, p. 173. — Williams, R. E.: J. Inorg. Nucl. Chem. *20*, 198 (1961).

[172] Gaines, D. F., Iorns, T. V.: J. Am. Chem. Soc. *89*, 4249 (1967).

[173] — — J. Am. Chem. Soc. *90*, 6617 (1968).

H. D. Johnson, II, and S. G. Shore

[174] Burg, A. B., Heinen, H.: Inorg. Chem. *7*, 1021 (1968).
[175] Iorns, T. V., Gaines, D. F.: Abstracts, 157th Meeting of the American Chemical Society, Minneapolis, Minn., April, 1969.
[176] Johnson, H. D., II: Ph. D. Thesis, The Ohio State University, 1969.
[177] Dahl, L. F., Calabrese, J.: cited in reference 173.
[178] Geanangel, R. A.: Ph. D. Thesis, The Ohio State University, 1968.
[179] — Johnson, H. D., II, Shore, S. G.: Unpublished results.
[180] Marynick, D., Onak, T.: J. Chem. Soc. A 1797 (1969).
[181] Shapiro, I., Landesman, H.: J. Chem. Phys. *33*, 1590 (1960).
[182] Figgis, B., Williams, R. L.: Spectrochim. Acta. *15*, 331 (1959).
[183] Hall, L. H , Subbanna, V. V., Koski, W. S.: J. Am. Chem. Soc. *86*, 3969 (1964).
[184] Burg, A. B.: J. Am. Chem. Soc. *90*, 1407 (1968).
[185] Gaines, D. F.: J. Am. Chem. Soc. *88*, 4528 (1966).
[186] Burg, A. B., Sandhu, J. S.: J. Am. Chem. Soc. *87*, 3787 (1965).
[187] Onak, T., Dunks, G. B.: Inorg. Chem. *3*, 1060 (1964).
[188] Gaines, D. F., Martens, J. A.: Inorg. Chem. *7*, 704 (1968).
[189] — J. Am. Chem. Soc. *91*, 1230 (1969).
[190] — Iorns, T. V.: Inorg. Chem. *7*, 1041 (1968).
[191] Reference 14 and references cited therein.
[192] Hough, W. V.: U.S. Patent 3,167,559.
[193] Raver, H. R., Bryher, H. B., Soborovsky, Z.: Zh. Obshch. Khim. *38*, 1328 (1968).
[194] Onak, T., Friedman, L. B., Hartsuck, J. H., Lipscomb, W. N.: J. Am. Chem. Soc. *88*, 3439 (1966).
[195] Friedman, L. B., Lipscomb, W. N.: Inorg. Chem. *5*, 1752 (1966).
[196] Hough, W. V., Edwards, L. F., Stang, A. J.: J. Am. Chem. Soc. *85*, 831 (1963)
[197] Tucker, P. M., Onak, T.: 158th National Meeting of the American Chemical Society, New York, New York, 1969, Inorganic Division Paper No. 105.
[198] Allerhand, A., Odom, J. D., Moll, R. E.: J. Chem. Phys. *50*, 5037 (1969).
[199] Tucker, P. M., Onak, T.: J. Am. Chem. Soc. *91*, 6869 (1969).
[200] Grimes, R., Wang, F. E., Lewin, R., Lipscomb, W. N.: Proc. Natl. Acad. Sci. U.S. *47*, 996 (1961).
[201] Hall, L. H., Subbanna, V. V., Koski, W. S.: J. Am. Chem. Soc. *86*, 3969 (1964).
[202] Gunn, S. R., Kindsvater, J. H.: J. Phys. Chem. *70*, 1114 (1966).
[203] Pinson, J. W., Lin, G.: Abstracts, 158th National Meeting of the American Chemical Society, New York, New York, 1969, Inorganic Division Paper No. 98.
[204] Grimes, R., Lipscomb W. N.: Proc. Natl. Acad. Sci. U.S. *48*, 496 (1962).
[205] Williams, R. E., Gerhart, F. J., Pier, E.: Inorg. Chem. *4*, 1239 (1965).
[206] — Gibbons, S. G., Shapiro, I.: J. Chem. Phys. *30*, 320 (1959).
[207] More, E. B., Lohr, L. L., Lipscomb, W. N.: J. Chem. Phys. *27*, 209 (1957).
[208] Guter, G. A., Schaeffer, G. W.: J. Am. Chem. Soc. *78*, 3546 (1956). — Hawthorne, M. F., Miller, J. J.: J. Am. Chem. Soc. *80*, 754 (1958). — Miller, J. J., Hawthorne, M. F.: J. Am. Chem. Soc. *81*, 4501 (1959). — Hough, W. V., Edwards, L. J.: Advan. Chem. Ser. *32*, 192 (1961).
[209] Carter, J. C., Mock, N. L. H.: J. Am. Chem. Soc. *91*, 5891 (1969).
[210] Palchak, R. J.: unpublished results. Cited in: Production of Boranes, p. 267; Holtzman, R. T., Editor. New York: Academic Press 1967.
[211] Williams, R. E., Gerhart, F. J.: J. Am. Chem. Soc. *87*, 3513 (1965).
[212] Onak, T. P., Williams, R. E.: Inorg. Chem. *1*, 106 (1962).
[213] Hawthorne, M. F., Pitochelli, A. R., Strahm, R. D., Miller, J. J.: J. Am. Chem. Soc. *82*, 1825 (1960).
[214] Lipscomb, W. N.: J. Chem. Phys. *28*, 170 (1958).

144

[215] Gaines, D. F., Schaeffer, R.: Proc. Chem. Soc., 267 (1963).

[216] Lutz, C. A., Phillips, D. A., Ritter, D. M.: Inorg. Chem. *3*, 1191 (1964).

[217] Gunn, S. R., Kindsvater, J. H.: J. Phys. Chem. *70*, 1114 (1966).

[218] Enrione, R. E., Boer, F. P., Lipscomb, W. N.: J. Am. Chem. Soc. *86*, 1451 (1964). — Enrione, R. E., Boer, F. P., Lipscomb, W. N.: Inorg. Chem. *3*, 1659 (1964).

[219] Dobson, J., Keller, P. C., Schaeffer, R.: Inorg. Chem. *7*, 399 (1968).

[220] Benjamin, L. E., Stafiej, S. F., Takacs, E. A.: J. Am. Chem. Soc. *85*, 2674 (1963); also see footnote 7 in reference 219.

[221] Maruca, R., Odom, J. D., Schaeffer, R.: Inorg. Chem. *7*, 412 (1968); the reaction using normal B_2H_6 is reported in reference 223.

[222] Ditter, J. F., Spielman, J. R., Williams, R. E.: Inorg. Chem. *5*, 118 (1966).

[223] Dobson, J., Schaeffer, R.: Inorg. Chem. *7*, 402 (1968).

[224] Keller, P. C.: J. Am. Chem. Soc. *91*, 1231 (1969).

[225] — Abstracts, 158th National Meeting of the American Chemical Society, New York, N. Y., 1969, Inorganic Division Paper No. 106.

[226] Centofanti, L. F., Parry, R. W.: Inorg. Chem. *7*, 1005 (1968).

Received January 26, 1970

Preparative Aspects of Boron-Nitrogen Ring Compounds

A. Meller

Institut für Anorganische Chemie, Technische Hochschule Wien, Austria

Contents

I.   Introduction.. 146

II. Cyclic Systems of Alternating B—N Units.......................... 147
     1. General 147
     2. Syntheses of the Borazine Ring............................... 148
     3. Substitution Reactions on Borazines 164
     4. Miscellaneous Borazine Derivatives and Reactions 168
     5. Borazine Complexes... 171
     6. s-Diazadiborines (—BR—NR'—)₂ 172
     7. s-Tetraza-tetraborines (—BR—NR'—)₄ 175

III. B—N Ring Systems Involving Annular Boron-Boron Bonds 176
     1. 2,5-Diaza-1,3,4-triborines 177
     2. 1,4-Diaza-2,3,5,6-tetraborines 178
     3. 1,3,5-Triaza-2,4,6,7-tetraborines 178

IV. B—N Ring Systems with Annular Nitrogen-Nitrogen Bonds 179
     1. 1,3,4-Triaza-2,5-diborines 179
     2. Cyclotetrazenoboranes.. 181
     3. 1,2,4,5-Tetraza-3,6-diborines 181
     4. 1,2,4,6-Tetraza-3,5,7-triborines 185

V.   References 185

I. Introduction

Several recent reviews have discussed the chemistry of boron-nitrogen compounds with particular emphasis on boron-nitrogen ring systems [1-7].

It is the purpose of this article to survey significant preparative methods leading to B—N ring compounds and to review the chemistry of such compounds. Discussion will be limited to those systems containing only B and N as annular atoms and in which the boron is three-coordinated. Special emphasis is placed on the more recent literature and the evaluation of current trends in this field of chemistry.

Several general methods are available for synthesizing systems with alternating boron and nitrogen atoms in the ring, but more specialized and refined procedures must be used to prepare cyclic B—N compounds with unsymmetrical distribution of the boron and nitrogen atoms in the ring.

Cyclic species of equal numbers of alternating boron and nitrogen atoms may be considered as condensation products of monomeric B—N units and recent work by Paetzold [8-11)] gives some credence to Wiberg's view that boronimides, XB≡NR, are possible intermediates in the formation of cyclic B—N compounds. Boronimides may polymerize or may be trapped by 1,3-dipolar agents [8-11)]. In similar fashion, the formation of the cyclotetrazenoborane ring system *1*

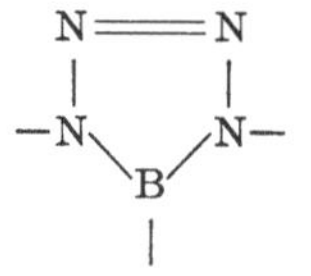

$$
\begin{array}{c}
N\!=\!\!=\!N \\
| \qquad | \\
-N \qquad N- \\
\diagdown \quad \diagup \\
B \\
|
\end{array}
$$

1

has been explained by cycloaddition of a boronimide to an organic azide [7)].

II. Cyclic Systems of Alternating B—N Units

1. General

The thermally most stable and best characterized cyclic boron-nitrogen systems containing alternating B—N groups are the symmetric six-membered borazine (s-triaza-triborine) *2* heterocycles. Other ring systems with symmetric arrangements of B and N atoms are the eight-membered s-tetrazatetraborine *3* systems and the four membered s-diazadiborines *4*. The formation of the latter is governed by steric factors.

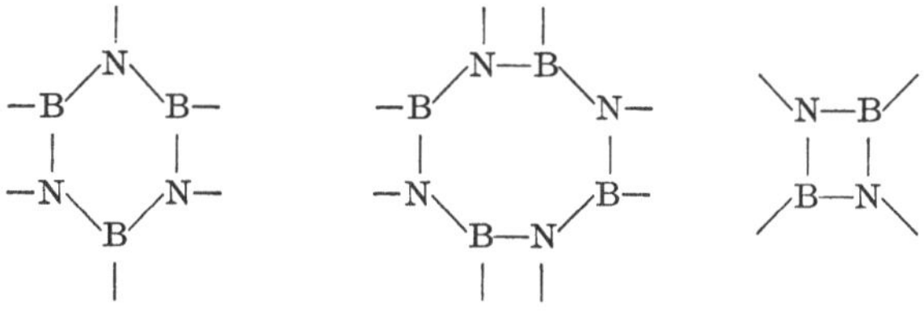

$$
\begin{array}{ccc}
2 & 3 & 4
\end{array}
$$

These induce difficulties in exocyclic substitution of the latter moieties; however, the hydrolytic stability of s-diazadiborines is superior to that of borazines in some instances. Systems containing N—N or B—B bonds in the ring appear to be thermally less stable than those which contain only B—N bonds. For other boron-nitrogen ring systems, an excellent survey [7] has been presented recently including a comprehensive tabulation of compounds. In the present article consideration of these other B—N rings is limited to the preparative aspects.

2. Syntheses of the Borazine Ring

2.1. Synthesis of Borazine and N-Alkyl-borazines

Among the synthetic procedures which have been elucidated since the first preparation of borazine by Stock and Pohland [13], the following two methods are of major significance:

a) The reaction of a metallic borohydride with ammonium chloride as depicted in Eq. (1), which provides

$$3 NH_4Cl + 3 M^IBH_4 \rightarrow [HNBH]_3 + 3 M^ICl + 9 H_2 \qquad (1)$$

a yield 35% of borazine when $NaBH_4$ is used in triglyme solution [14].

b) The reduction of 2,4,6-trihaloborazines with alkali metal hydroborates using donor molecules for absorption of the diborane which is formed simultaneously in this reaction. A carefully elaborated method using $NaBH_4$ and tri-n-butylamine for the reduction of B-trichlorborazines has been described in Inorganic Syntheses [15].

$$[HNBCl]_3 + 3 NaBH_4 + 3 (n-C_4H_9)_3N \rightarrow [HNBH]_3 + 3 NaCl + $$
$$+ 3 (n-C_4H_9)_3N:BH_3 \qquad (2)$$

The yield is over 40%.

Both of the above methods are equally effective in preparing N-alkyl or aryl-substituted borazines. Yields are usually higher for N-organoborazines than they are for the parent borazine. This is most probably due to the fact that during the preparation of borazine side reactions occur with the elimination of hydrogen from the N and B atoms; on the other hand, elimination of R—H (R = alkyl, aryl) is less favored. Intermediates postulated in the formation of the borazine ring are either the boronimide [HN≡BH] [12] or an aminoborane trimer (cyclotriborazane) $[H_2NBH_2]_3$ [16]. Recent investigations by Shore [5] have shown that cyclotriborazane is not obtained by pyrolysis of amine-borane H_3NBH_3. However, methylamine-borane $CH_3NH_2BH_3$ gives, at 100°, 1,3,5-trimethylcyclotriborazane in nearly quantitative yield and N-trimethylborazine results from the latter at 200° [17].

$$3\ CH_3NH_2BH_3 \xrightarrow[100°]{-3\ H_2}$$

$$\xrightarrow[200°]{-3\ H_2}$$

(3)

Hence different mechanistic routes appear feasible for the formation of borazine and its N-alkylated derivatives. That route involving the cyclization of an aminoborane to a cyclotriborazane will involve fewer side reactions than that proceeding via a boronimide. Recent studies [18] suggest, that a six-membered linear B—N chain could be the first intermediate and that borazine will result from an intermolecular dehydrogenation and ring closure to yield first a cyclohexene-analogue which rapidly dehydrogenates to produce the borazine. Synthetic and trapping methods in the study of intermediates of 1,3,5-trimethylcyclotriborazane [19] arrive at the same conclusion. These intermediates, however, were obtained with an amine-hydrochloride as a reactant and do not give conclusive evidence against the intermediate formation of a boronimide species as postulated by Paetzold [8-11].

In addition to information compiled elsewhere [1-7] the preparation of N-trimethylborazine has been described in Inorganic Syntheses [20] and the standard procedures have been used to make several isotopically labeled species of borazine [21] and N-methylborazine [22,23]. Recently, the direct synthesis of N-organoborazines by the interaction of nitriles with diborane or alkali metal hydroborates, rather than the two-step procedures involving the reduction of the corresponding B-chloroborazines, has received much attention. Besides the extensive study by Wade et al. [24], it has been shown [25] that borazines can be prepared in high yield according to the reaction depicted in [Eq. (4)].

$$12\ RCN + 9\ NaBH_4 + 12\ BF_3 \cdot THF \xrightarrow[THF]{} 4\ [RCH_2NBH]_3 + 9\ NaBF_4 + 12\ THF \quad (4)$$

Hydroboration of nitriles has also been used in syntheses of N-halo-organoborazines (see II.2.4.). Furthermore, a hydrogen-transfer reaction has been described by which borazines are formed through the reaction of trimeric aminoboranes with dimethylaminoborane in high yield [26] [Eq. (5)]. The latter method, however, is primarily of theoretical interest.

$$3\ (CH_3)_2NBH_2 + [RNHBH_2]_3 \rightarrow 3\ (CH_3)_2NHBH_3 + [RNBH]_3 \quad (5)$$

149

2.2. Syntheses of 2,4,6-Trihalo- and 1,3,5-Trisorgano-2,4,6-trihaloborazines

As is the case for the synthesis of B—H compounds, N-organosubstituted 2,4,6-trihaloborazines are obtained in better yield and higher purity than the corresponding N—H borazines. The major reason for this occurrence appears to be a tendency for intermolecular condensation. Hydrogen halide is absorbed by other decomposition products and therefore cannot be utilized as a measure of decomposition of the stored (or pyrolized) material.

$$2 \quad \begin{array}{c} \text{H} \\ \text{N} \\ \text{XB} \quad \text{BX} \\ \text{HN} \quad \text{NH} \\ \text{B} \\ \text{X} \end{array} \longrightarrow \begin{array}{c} \text{H} \qquad \text{X} \\ \text{N} \qquad \text{B} \\ \text{XB} \quad \text{B} \text{—} \text{N} \quad \text{NH} \\ \text{HN} \quad \text{NH} \quad \text{XB} \quad \text{BX} \\ \text{B} \qquad \text{N} \\ \text{X} \qquad \text{H} \end{array} + \text{ HX etc.} \qquad (6)$$

The tendency to undergo a condensation reaction is clearly pronounced in solution and can be observed with virtually all types of N—H borazines if the B-substituent is a suitable partner for ionic reactions. For example, B-triphenylsiloxyborazines condense in solution with the formation of hexaphenyldisiloxane, water, and condensed B—N ring systems which in due course will react with the water [27]. The tendency to condense increases in the series $[HNBF]_3 < [HNBCl]_3 < [HNBBr]_3 < [HNBI]_3$. Under more stringent conditions H_2 and BF_3 are evolved rather than HF in the case of 2,4,6-trifluoroborazine [28]. $[HNBI]_3$ is extremely sensitive and the only known method of preparing it involves the interaction of triiodoborane with hexamethyldisilazane in solution [29,30].

$$3 \, [(CH_3)_3Si]_2NH + 3 \, BI_3 \xrightarrow{CCl_4} 6 \, (CH_3)_3SiI + [HNBI]_3 \qquad (7)$$

Preparation of B-fluoroborazines cannot be effected from trifluoroborane and amines by a Brown-Laubengayer type of reaction. However, direct synthesis of B-fluoroborazine rings has been reported recently and involves the dehydrofluorination of primary amine-trifluoroborane adducts with the aid of the adducts of sterically hindered tertiary amines with trifluoroborane. 2,4,6-Trifluoroborazines and ammoniumtetrafluoroborates [31] are obtained in excellent yield.

$$3 \, RNH_2 \cdot BF_3 + 6 \, R_3'N \cdot BF_3 \rightarrow [RNBF]_3 + 6 \, R_3'NH^+BF_4^- \qquad (8)$$

This reaction was studied for a large number of amines, yielding 2,4,6-trifluoroborazine as well as sterically hindered derivatives such as 1,3,5-tris(t.-butyl)-2,4,6-trifluoroborazine. The reaction proceeds smoothly when BF_3-etherate is added to the solution of a mixture of the primary amine and diisopropylethylamine in benzene or ether.

Most commercial samples of trichloroborazine contain condensation products which are partially insoluble in diethylether. Purification can be achieved by extraction of the 2,4,6-trichloroborazine with diethylether and sublimation of the pulverized extract in high vacuum below 65° under mechanical motion. The long time of refluxing in chlorobenzene necessary in the laboratory scale synthesis by the original Brown-Laubengayer method [32,33) (but cf. ref. [44)] reduces the yield of pure 2,4,6-trichloroborazine to 40% compared to 80% obtained in the analogous preparation of 1,3,5-trimethyl-2,4,6-trichloroborazine [34). Condensation of 2,4,6-trihaloborazine is faster in solution than in the solid state. The Grignard alkylation of 2,4,6-trihaloborazines in particular promotes intermolecular B—N condensation, probably due to an exchange reaction [35). The latter

$$\diagdown NH + RMgX \rightarrow \diagdown NMgX + RH; \quad \diagdown NMgX + XB \diagup \rightarrow \diagdown N-B \diagup + MgX_2 \quad (9)$$

competes with the normal alkylation process and leads to N—B linked B-alkyl-polyborazines [35-37).

$$+ \text{ polycondensation products} \qquad (10)$$

It is noteworthy that biphenyl-type borazines result as byproducts from such reactions in up to 25% yield, while yields of decamethyl-N—B-diborazyl [38) from 1,2,3,4,6-pentamethyl-5-lithioborazine [39,40) and 1,2,3,4,5-pentamethyl-6-chloroborazine range between 40 and 50%. The formation of naphthalene-type borazine condensation products 5 has been claimed to occur in such reactions [41,42).

5

However, mass-spectroscopic studies [37] could not confirm these findings. If a Brown-Laubengayer synthesis is performed utilizing BCl_3 and a mixture of ammonium chloride and methyl-ammonium chloride [43] separation by distillation does not yield 2,4,6-trichloroborazines substituted unsymmetrically at the nitrogen, i.e. $H_2(CH_3)N_3B_3Cl_3$ and $H(CH_3)_2N_3B_3Cl_3$. This event is due to the ready condensation of B-haloborazines.

However, 1,2,4,6-tetramethyl- and 1,2,3,4,6-pentamethylborazine, which will not condensate during distillation, were obtained by Grignard methylation of the raw material resulting from an "unsymmetrical Brown-Laubengayer synthesis" [36].

Recently, a method has been described which eliminates the long reflux time needed for completion of the Brown-Laubengayer reaction. Only 2—3 hours refluxing are required for the formation of 2,4,6-trichloroborazine from the interaction of NH_4Cl with the acetonitrile-trichloroborane adduct [44]. The yield is about 60%. No mechanism has been proposed, but it is known that in other B—N ring syntheses (cf. Sections II.2.4. and II.7) the use of trihaloborane complexes is an essential feature.

N-Trialkyl-B-trihaloborazines can be prepared without difficulty from alkylammonium halide or alkylamine and trihaloborane [4] when the resulting alkylammonium tetrahaloborate or alkylamine-trihaloborane is refluxed in chlorobenzene. Yields are nearly quantitative in most cases. This same observation holds true for N-triaryl-B-trihaloborazines. Extensive studies using thermogravimetric methods in addition to preparative methods did not provide for the isolation or identification of any intermediates between the amine-trihaloborane and the borazine stage of the reaction [45—49].

$$RNH_3^+BCl_4^- \xrightarrow{-\ HCl} RNH_2BCl_3 \xrightarrow{-\ 2\ HCl} \tfrac{1}{3}[RNBCl]_3 \qquad (11)$$

The borazine formation from the alkylamine-trihaloborane is influenced by the reaction temperature; for example, in boiling benzene no borazine is obtained because the temperature is not high enough. In most cases, however, a borazine can be prepared even without heating if triethylamine is added to the reaction mixture [50], though, this procedure does not present any practical advantage in borazine ring synthesis. In those cases where triethylamine cannot be avoided (i.e. in the synthesis of N-pentafluoroaryl-B-haloborazines or of the s-tetrazatetraborines) [cf. Sections II.2.4., II.7], the triethylamine-trihaloborane complex must be used as a starting material; mere addition of the base to the reaction mixture will not yield the desired products. This event might indicate that the aminoborane intermediate (which can be isolated by thermal decomposition of the amine-borane complex in the "sterically hindered" case [51] is normally not an intermediate in the borazine formation. The rate of evolution of hydrogen chloride in thermal dehydrohalogenation of prim. organoamine-trihaloboranes is not only a function of the organic groups but also of the solvent used. Fractional order evolution of hydrogen chloride, however, might suggest an initial equilibrium state or chain reaction [51].

In the synthesis of N-trialkyl-B-trihaloborazines no condensation side reactions are observed. Elimination of RX does not occur under the conditions normally applied in synthesis or substitution reactions. Even N-trialkyl-B-triiodoborazines can be made from the corresponding alkyl-ammonium tetraiodoborates or from triethylamine-triiodoborane and alkylamine [46].

2.3. Syntheses of Hexachloroborazine

Elements other than hydrogen, lithium, boron, carbon and silicon were virtually unknown as substituents on the borazine N-atoms. Fairly recently, Paetzold [52] obtained hexachloroborazine by thermal decomposition of $(Cl_2BN_3)_3$. It may be noted that N—N substituted borazines have not yet been isolated, though attempts have been made to prepare such materials [53]. However, formation of tetrakis(diethylamino)-1,3,2,4-diazadiborine was accomplished by photolytic decomposition of bis (diethylamino)azidoborane [54] in solution as shown in Eq. (12).

$$2\ [(C_2H_5)_2N]_2BN_3\ \rightarrow\ 2\ N_2\ +\ (C_2H_5)_2N-N\underset{\underset{\displaystyle N(C_2H_5)_2}{|}}{\overset{\overset{\displaystyle N(C_2H_5)_2}{|}}{\underset{B}{\overset{B}{\diamond}}}}N-N(C_2H_5)_2 \qquad (12)$$

Experiments designed to prepare borazines with exocyclic bonding to oxygen from the ring nitrogen led to violent explosions, whether or not the preparation was attempted by using hydroxylammonium chloride and trichloroborane in a Brown-Laubengayer synthesis [55], or the following procedure was utilized [56].

$$CH_3ONH_3Cl + LiBH_4\ \rightarrow\ CH_3ONH_2BH_3\ \not\rightarrow\ [CH_3ONBH]_3 \qquad (13)$$

In the latter reaction, a colorless extremely brisant liquid was obtained. On the basis of infrared spectral data structure 6 was assigned to the product.

$$\begin{array}{c}
\text{H} \qquad \text{OCH}_3 \\
\text{H}-\text{B} \quad \text{N} \quad \text{B}-\text{H} \\
\text{H} \qquad \qquad \text{H} \\
\text{CH}_3\text{O} \qquad \text{OCH}_3 \\
\text{H}-\text{N} \quad \text{N}-\text{H} \\
\text{B} \\
\text{H} \qquad \text{H}
\end{array}$$

6

Two additional procedures have been developed for the synthesis of hexachloroborazine [Eq. (14) and (15)].

$$3\,\text{BCl}_3 + 3\,(\text{CH}_3)_3\text{Si}-\text{N}-\text{Si}(\text{CH}_3)_3 \xrightarrow[-78°]{\text{CH}_2\text{Cl}_2} [\text{ClNBCl}]_3 + 6\,(\text{CH}_3)_3\text{SiCl} \quad (14)$$
$$\overset{|}{\text{Cl}}$$

The mechanism of reaction according to Eq. (14) is discussed by the authors [57]. They postulate the formation of the adduct $(\text{R}_3\text{Si})_2\text{N}(\text{Cl})\text{BCl}_3$ which is followed by rupture of the Si—N bonds resulting in the formation of halosilanes. A similar reaction has been reported elsewhere (see ref. [58] and lit. cited therein). However, in the reaction of N-chlorohexamethyldisilazane with trichloroborane substantial side reactions seem to occur. Several liquid products have not yet been identified.

Despite its utilization of the dangerous NCl_3, the most advantageous method appears to be the reaction illustrated in Eq. (15). This reaction results in the formation of hexachloroborazine in nearly quantitative yield [59].

$$3\,\text{NCl}_3 + 3\,\text{BCl}_3 \rightarrow 3\,\text{Cl}_3\text{NBCl}_3 \xrightarrow[\text{max. } 45°, \text{ N}_2]{\text{CCl}_4} [\text{ClNBCl}]_3 + 3\,\text{Cl}_2 \quad (15)$$

It is worth noting that the adduct of Cl_3N with BCl_3 can be obtained only at low temperature [60]. Hexachloroborazine reacts with water to boric acid and chloramine; with methanol, trimethoxyborane and methylhypochlorite are produced and, with ethanol, triethoxyborane and acetaldehyde are obtained [59]. Substitution reactions on the B—N ring of hexachloroborazine have not yet been studied.

2.4. Syntheses of Haloalkylborazines

The chemistry of borazines in which either the nitrogen or boron atoms of the ring are substituted with haloalkyl groups has been developed within the last few years only. The lone haloalkylborazine known before 1964 was 2,4,6-tris(chloromethyl)borazine, obtained from the interaction of 2,4,6-trichloroborazine with diazomethane [61]. The reaction of heptafluoropropyl-lithium with 1,3,5-triphenyl-2,4,6-trichloroborazine does not yield the desired heptafluoropropylborazine derivative [62]. The first identified B-fluoroorganoborazine was 2,4,6-tris(pentafluorophenyl)-borazine, obtained from the interaction of 2,4,6-trichloroborazine with pentafluorophenyllithium [63]. This compound, like other B-fluoroaryl-borazines containing organic substituents at the nitrogen sites [64], is an extremely stable substance. In contrast, the reaction of B-chloroborazines with β-fluoroalkyl Grignard derivatives yields B-fluoroborazines, [RNBF]₃, rather than B-fluoroalkylborazines [64]. The first reported N-haloalkylborazines encompass 1,3,5-tris-(β-trichloroethyl)- and 1,3,5-tris(β-trifluoroethyl)borazine. These compounds were obtained from the interaction of the corresponding halogenated acetonitriles with diborane in etheric solvents [65,66]. Isolation of an iminoborane intermediate, $Cl_3CCH=NBH_2$, leads to the assumption that iminoboranes of halogenated acetonitriles might be more stable than those of other nitriles. This observation has recently been confirmed by the preparation of highly halogenated iminoboranes by the reaction of trihaloboranes with a variety of nitriles having electron-attracting substituents (see ref. [67] and [68] and lit. cited therein). On preparation of 1,3,5-tris(β-trifluoro-ethyl)-2,4,6-trichloroborazine by a Brown-Laubengayer synthesis [Eq. (16)] the completion of the reaction (to yield the desired borazine) requires refluxing in chlorobenzene for 60 hrs.

$$3\ CF_3CH_2NH_3Cl + 3\ BCl_3 \xrightarrow[\text{C}_6\text{H}_5\text{Cl}]{\text{60 hrs., 132°}} [CF_3CH_2NBCl]_3 + 9\ HCl \qquad (16)$$

The same reaction, originating from the non-fluorinated compound (i.e. ethylammonium chloride) yields the borazine after refluxing for only 15 to 20 hours in the same solvent. The postulated formation of $[CF_3CH_2NBCl]_3$ [69] could not be confirmed [64]. However, 2,4,6-organo-and fluoroorgano-substituted derivatives of 1,3,5-tris(trifluoroethyl) borazine have been made from $[CF_3CH_2NBCl]_3$ by Grignardalkyla-tions [64].

On reaction of pentafluoroaniline with tribromoborane at 70°,

$$3\ C_6F_5NH_2 + 3\ BBr_3 \longrightarrow 3\ C_6F_5NH_2BBr_3 \xrightarrow{70°} 3\ C_6F_5NHBBr_2 \qquad (17)$$

(pentafluoroanilino)dibromoborane is formed [70]. However, pure 1,3,5-tris-(pentafluorphenyl)-2,4,6-tribromoborazine cannot be obtained by thermal treatment of the aminoborane. 1,3,5-Tris(pentafluorphenyl)-2,4,6-trihaloborazines have been obtained in good yield and high purity by the following reaction sequence [71,72] [Eq. (18)].

$$3 \ (C_2H_5)_3N + 3 \ BX_3 \rightarrow 3 \ (C_2H_5)_3N : BX_3$$

$$3 \ (C_2H_5)_3N : BX_3 + 3 \ (C_2H_5)_3N + 3 \ C_6F_5NH_2 \rightarrow$$

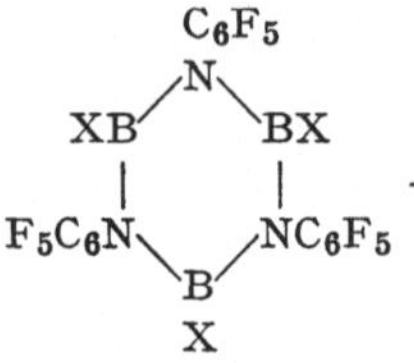

$$+ 6 \ (C_2H_5)_3NHCl$$

(18)

It is essential in this preparation to start from the preformed triethylamine-trihaloborane complex. Addition of triethylamine as a hydrogen halide acceptor at a later stage of the reaction (i.e. to the pentafluotoanilino-dihaloborane) does not yield the N-tris(perfluoroaryl) compound. Rather, a mixture of products is obtained from which the desired compound cannot be isolated in a pure state. Fluoroorganoborazine derivatives described so far are listed in Table 1. The table also includes pentafluorophenoxy- and pentafluoroanilinoborazines.

The above data suggest a different mechanism for the reaction originating from triethylamine-trihaloborane and pentafluoroaniline as compared to the reaction between pentafluoroaniline and trihaloborane. However, it was shown that N-tris(pentafluorophenyl)-B-trihaloborazines can be prepared in the Brown-Laubengayer fashion in n-nonane (b.p. 152°) by refluxing for 60 hours [73].

N-tris(γ-fluoroorgano)borazine compounds can be made either by the standard Brown-Laubengayer method or by interacting the corresponding nitriles with diborane in dimethoxyethane. Several N-tris(pentafluorobenzyl)borazine derivatives have been described [74].

N-Fluoroorganoborazines have been subjected to substitution reactions. These lead to various N-tris(fluoroorgano)-B-halogeno- [72-74], -B-pseudohalogeno- [72], -B-amino- [72,74], -B-fluoroorgano [64,72-74] and B-organoborazines [64,72]. These compounds were prepared by standard substitution procedures. Hexakis(perfluoroorgano)borazines and trisaryl-tris(hexafluoroorgano)borazines exhibit high thermal stability [72,73] as well as relatively high resistance towards hydrolysis [73,74].

Unsymmetrically substituted perfluorovinylborazines of high thermal stability have been reported [75] recently. They were obtained by the reaction of trifluorovinyllithium with 1,2,3,5-tetramethylborazine and 1,2,3,4,5-pentamethylborazine respectively.

Table 1

Compound	mp. °C	Subl./Dist. temp./mm	Ref.
$[CF_3CH_2NBH]_3$	40—41	106—111/92	65,66)
$[CF_3CH_2NBCl]_3$	33—35	112—114/9	64)
$[CF_3CH_2NBNH_2]_3$	157	100/10 *)	74)
$[CF_3CH_2NBN(CH_3)_2]_3$		138/10 *)	74)
$[CF_3CH_2NBC_6H_5]_3$	140	251/9 *)	64)
$[CF_3CH_2NBC_6F_5]_3$	185	150/0,03 *)	64)
$[HNBC_6F_5]_3$	211—213	200	63)
$[CH_3NBC_6F_5]_3$	275	245/9 *)	64)
$[CH_3NBCF=CF_2]_3$			75)
$[C_6H_5NBC_6F_5]_3$	340	308/9 *)	64)
$[HNBCH_2C_6F_5]_3$	98—103	194/0.001 *)	74)
$[CH_3NBCH_2C_6F_5]_3$	118—122	206/0.001 *)	74)
$[C_6H_5NBCH_2C_6F_5]_3$	189—192	239/0.001 *)	74)
$[C_6F_5NBF]_3$	190	148/0.001 *)	72,73)
$[C_6F_5NBCl]_3$	254	185/0.001 *)	71,73)
$[C_6F_5NBBr]_3$	258—260	188/0.001 *)	72,73)
$[C_6F_5NBC_6H_5]_3$	376	284/0.001 *)	72,73)
$[C_6F_5NBC_6F_5]_3$	402—405	250/0.001 *)	72)
$[C_6F_5NBCH_2C_6F_5]_3$	185—188	212/0.001 *)	74)
$[C_6F_5NBNCO]_3$	156—158		72)
$[C_6F_5NBNCS]_3$	219—221		72)
$[C_6F_5NBNH_2]_3$	259	202/0.001 *)	74)
$[C_6F_5CH_2NBH]_3$	116—118	168/0.001 *)	74)
$[C_6F_5CH_2NBF]_3$	107—109	170/0.001 *)	74)
$[C_6F_5CH_2NBCl]_3$	144—148	185/0.001 *)	74)
$[C_6F_5CH_2NBCH_2C_6F_5]_3$	206—210	242/0.001 *)	74)
$[C_6F_5NBNHC_6H_5]_3$	240—243	236/0.001 *)	72)
$[HNBOC_6F_5)_3$	156	165/0.001 *)	72)
$[CH_3NBOC_6F_5]_3$	115	160/0.001 *)	72)
$[C_6H_5NBOC_6F_5]_3$	221	222/0.001 *)	72)
$[CH_3NBNHC_6F_5]_3$	208	208/0.001 *)	72)
$(CH_3)_3N_3B_3(CH_3)_2CF=CF_2$		150/0.03	75)
$(C_6H_5)_3N_3B_3(H)_2CH_2CH_2CF_3$	128—130		69)
$[(CH_3)_3N_3B_3(C_4H_9)_2]_2(NHC_6F_4)_2$	201	294/0.001 *)	72)
$[(CH_3)_3N_3B_3(C_4H_9)]_n[(NHC_6F_4)_2]_n$	300		72)

*) Air bath-temperature

The greater stability of perfluoro-unsaturated B-compounds is not merely a matter of B—C π-bonding. This view [76] is supported by recent data [64,75]. In contrast to B-fluoralkylborazines, the corresponding chloroalkylborazines could be expected to be stable substances if one considers [HNBCH$_2$Cl]$_3$ [61]. Indeed, a mixture of B-chloromethyl-borazines can be obtained by chlorination of 2,4,6-trimethylborazine in CCl$_4$ by u.v. radiation. From this mixture pure [HNBCHCl$_2$]$_3$, which is probably the most stable representative of these compounds, can be isolated [77]. Prolonged chlorination, however, leads to [HNBCCl$_3$]$_3$ and finally to [HNBCl]$_3$ and hexachloroethane [27,77]. On the other hand, the chlorination of [CH$_3$NBCl]$_3$ does not yield a N-chloromethylborazine compound; rather the dimeric iminoborane [Cl$_2$C=N—BCl$_2$]$_2$ [77,78] is obtained besides unidentified products.

2.5. Sterically Hindered Borazines

One of the major problems in the chemistry of three-coordinated boron-nitrogen compounds is their relative instability towards hydrolysis [79,80]. This feature reduces their possible applications in many areas such as in the medicinal field or utilization for polycondensation-products. However, steric hindrance by bulky substituents on the borazine ring leads to compounds with enhanced hydrolytic stability, particularly when no solvent is present in which they may dissolve. This observation has resulted in extensive research on sterically hindered boron-nitrogen ring compounds.

There are only two known borazine compounds having t-butyl groups on the nitrogen atoms. 1,3,5-Tris-tert.-butylborazine was made by thermal decomposition of t-butylamine-borane [81] [Eq. (19)].

$$3\ t\text{-}C_4H_9NH_2BH_3 \xrightarrow{360°} \begin{matrix} & C_4H_9\text{-}t \\ & N \\ HB & BH \\ | & | \\ t\text{-}C_4H_9\text{—}N & N\text{—}C_4H_9\text{-}t \\ & B \\ & H \end{matrix} +\ 6\ H_2 \qquad (19)$$

This same substance has also been obtained from triethylamine-borane and t-butylamine [82] or from sodium borohydride, dioxane-trifluoroborane and t-butylamine respectively [83]. It is not attacked by boiling water, though it slowly hydrolyzes in boiling aqueous acetone. All attempts to accomplish substitution on the B-atoms were unsuccessful [81,84]. Most probably, 1,3,5-tris(t.butyl)-2,4,6-trifluoroborazine, which was only recently prepared [31], will be rather unreactive. All other known B—N ring compounds with t-butyl groups at the N, exhibit either four- or eight-membered ring structures (cf. Sections II.6, II.7). This fact indicates that, at least in these B—N ring compounds, steric factors are predominant in determining the ring size. In the case of bis- and tris-(aminoboranes) other factors, such as the basicity of the amine, have been invoked in order to explain the degree of association [20]. Indeed, by application of chemical dehydrochlorination with triethylamine, it was shown that amine-trihaloboranes derived from primary amines will not yield borazines if steric hindrance is too severe [46,51]. Rather, tetrameric borazynes or, in some cases, diborylamine compounds or bis(amino)boranes are formed.

In borazine compounds having branched alkyl groups (other than t-butyl) at the N-atoms, a stepwise substitution of the (B)—H atoms in Grignard alkylation [84] is observed rather than the statistical substitution which occurs on alkylation of N-methylborazines with n-alkyl Grignard reagents [34]. It has been reported [49,85] that base dehydrochlorination of isobutylammonium trichlorophenylborate as well as that of isobutyl-amine-dichlorophenylborane results in a mixture of the trimeric and tetrameric N-isobutyl-B-phenylborazines [Eq. (20)].

$$7 \text{ i-C}_4\text{H}_9\text{NH}_2 \cdot \text{C}_6\text{H}_5\text{BCl}_2 + 14 \text{ (C}_2\text{H}_5\text{)}_3\text{N} \rightarrow [\text{i-C}_4\text{H}_9\text{NBC}_6\text{H}_5]_3 +$$
$$+ [\text{i-C}_4\text{H}_9\text{NBC}_6\text{H}_5]_4 + 14 \text{ (C}_2\text{H}_5\text{)}_3\text{NHCl} \tag{20}$$

The tetramer is said to be irreversibly converted into the trimeric product by heating above its melting point. In contrast to an earlier report claiming the synthesis of linear N-isobutyl-B-phenyl-polybora-zynes [86] only the normal borazine-formation could be observed on the pyrolysis of bis(isobutylamino)phenylborane [87]:

$$3 \text{ C}_6\text{H}_5\text{B(NH i-C}_4\text{H}_9\text{)}_2 \xrightarrow{260-315°} [\text{i-C}_4\text{H}_9\text{NBC}_6\text{H}_5]_3 + 3 \text{ i-C}_4\text{H}_9\text{NH}_2 \tag{21}$$

In these studies [87,88] gas chromatographic methods have been shown to be useful tools for the separation of hexaorganoborazines.

Steric hindrance in aryl-substituted borazines depends on o-substitution of the aryl groups. Significant results concerning the reaction of primary arylamines with trihaloboranes, have recently been discussed in detail [89,90]. While pyrolysis of aniline-trihaloboranes or o-toluidine-trihaloboranes yields the corresponding B-trichloroborazines, the base-promoted dehydrohalogenation of the latter results in the formation of a 1,3-diazaro-2,4-dibora-naphthalene compound besides a minor amount of borazine.

Hence, pyrolysis and base dehydrohalogenation must follow different paths in the cited reactions and probably in any other case involving steric or particular electron effects.

$$n \ o\text{-}C_6H_4(CH_3)NH_2 \cdot BCl_3 + 2n \ (C_2H_5)_3N \longrightarrow$$

$$+ \ (C_2H_5)_3NHCl + other \ products \tag{22}$$

When the aminoboranes obtained by the thermal dehydrohalogenation of 2,6-disubstituted anilines are treated with a tertiary base, dichloroborazines are obtained besides a diborylamine [Eq. (23)].

$$+ \ RNHB\text{-}N\text{-}B\text{-}NHR \qquad (23)$$

$X = $ alkyl, Cl, Br
$R = 2,6 \ C_6H_3(X)_2$

An interpretation for the apparent reduction at one of the B-atoms has not yet been presented.

B-2,6-Dimethylarylborazines have not been obtained in a direct borazine synthesis, though they can be prepared by Grignard arylation of 1,3,5-trimethyl-2,4,6-trichloroborazine [91,92]. These compounds show high hydrolytic stability, even in acid and basic medium, and they can be subjected to substitution reactions at the aromatic sites [91]. For example, free radical bromination [93] or Friedel-Crafts acetylation [94] has been accomplished.

2.6. Syntheses of Unsymmetrically Substituted Borazine Rings

Direct synthesis of unsymmetrical substituted borazines has been reported infrequently. The feature of borazines with unsymmetrical substitution at the boron atoms resides in the fact that 2-organo- or 2,4-diorgano-borazines may be utilized in polycondensation reactions to yield insoluble compounds of macroscopic linear structure.

In principle, the synthesis of unsymmetrically substituted moieties of borazine may be achieved by the same reactions which are suitable for preparing the symmetrical compounds.

For example, mixtures of ammonia and prim. organoamines (or their salts) can be reacted with diborane [95], a tetrahydroborate [96,97], or haloboranes [36] to yield N-unsymmetrically substituted borazines besides symmetrical materials. Separation of the products often presents major obstacles, particularly if boron-halogenated derivatives are obtained, since subsequent condensation reactions may render isolation of the products difficult. This difficulty, however, may be overcome by alkylation at B-atoms [36]. Gas chromatography has been used successfully to isolate products [98]. Borazines with unsymmetrical substitution at the boron atoms are obtained by reacting ammonia or amines with mixtures of alkyldiboranes [99]. However, for the preparation of B-unsymmetrical borazines, substitution reactions on preformed borazine-rings appear to be the preferential method. Recently, synthesis of B-mono- and B-dimethylborazine has been described by alkylation of borazine [HNBH]$_3$ with CH_3MgJ [97]. 1,3,5-Tris(2,6-disubstituted aryl)-2,4-dihaloborazines obtained in a borazine ring synthesis [89] were mentioned in the preceding chapter.

Unsymmetrically substituted tetra- and penta-n-butyl borazines are obtained by thermal decomposition of dibutylazidoborane [100] with the elimination of butene. Because of the hazards of handling boron azides, this method cannot be recommended.

2.7. Syntheses of Miscellaneous Borazine Compounds

Borazines with dialkylaminoalkyl [$R_2N(CH_2)_n-$] substituents on the nitrogen sites have been described in the patent literature. 1,3,5-Tris-(p-dimethylaminophenyl)-2,4,6-trialkylborazines are prepared either by the reaction of p-dimethylaminoaniline with BCl_3 followed by Grignard-alkylation [101], or by reaction of alkyldichloroboranes with p-dimethyl-aminoaniline [102]. The preparation of 1,3,5-tris(aminoalkyl)borazines by the reaction [103], is illustrated by Eq. (24).

$$9\ NaBH_4 + 12\ BF_3 \cdot THF + 12\ H_2N-(CH_2)_3N(CH_3)_2 \rightarrow$$
$$4\ [(CH_3)_2N(CH_2)_3NBH]_3 + 9\ NaBF_4 + 12\ THF + 24\ H_2 \tag{24}$$

It results in the formation of the corresponding borazinyl-N-alkyl-amine-boranes [$BH_3(CH_3)_2N(CH_2)_3NBH]_3$ when the reaction is carried out with only 6 molar equivalents of the diamine. Aromatic diamines have also been used in this synthesis [103]. Borazines with N-organosilicon substituents have been obtained by reacting silicon-containing nitriles, for example, 4-cyano-2,2,6,6-tetramethyl-2,6-disilatetrahydropyrane, with $NaBH_4$ and BF_3 in tetrahydrofuran.

N-Silylborazines form a class of compounds of which few representatives are known (see Table 2). Lately several new derivatives have been obtained by thermal decomposition of bis(trimethylsilyl)aminoborane compounds [105] as depicted in Eq. (25).

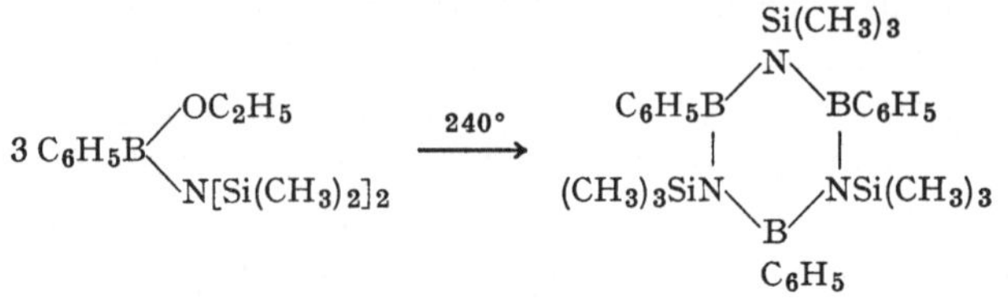

$$
3\ C_6H_5B \underset{N[Si(CH_3)_2]_2}{\overset{OC_2H_5}{\big<}} \xrightarrow{240°} \quad
\begin{array}{c}
Si(CH_3)_3 \\
N \\
C_6H_5B \diagup \quad \diagdown BC_6H_5 \\
| \qquad\qquad | \\
(CH_3)_3SiN \diagdown \quad \diagup NSi(CH_3)_3 \\
B \\
C_6H_5
\end{array}
\tag{25}
$$

Table 2. *N-Silylborazine compounds*

Compound	mp. °C	Subl./Dist.	Temp./Torr	Ref.
[H₃SiNBF]₃		25		[107]
[(CH₃)₃SiNBH]₃		104/1		[30]
[(CH₃)₃SiNBF]₃				[108]
[(CH₃)₃SiNBCl]₃	140	96/0.2		[105,106]
[(CH₃)₃SiNBC₆H₅]₃	156	160/0.02		[105]

During the last five years several borazines have been described in which the borazine ring forms the center of organic or, in one case, of an inorganic ring system: on heating the salt $[Cl_3P=N-PCl_3]\,[BCl_4]$ in tetrachloroethane, elimination of HCl occurs and a product is formed which is thought to have the following structure [109] 7:

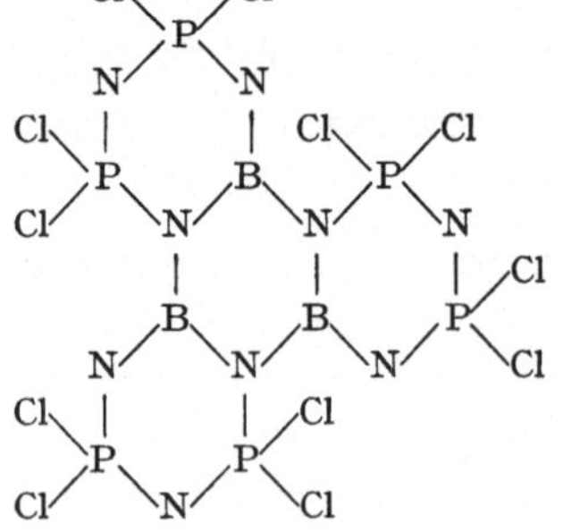

7

1,2,3,4,5,6-tris(2,2'-biphenylene)borazine *8* has been made by the following reaction [110,111] [Eq. (26)]

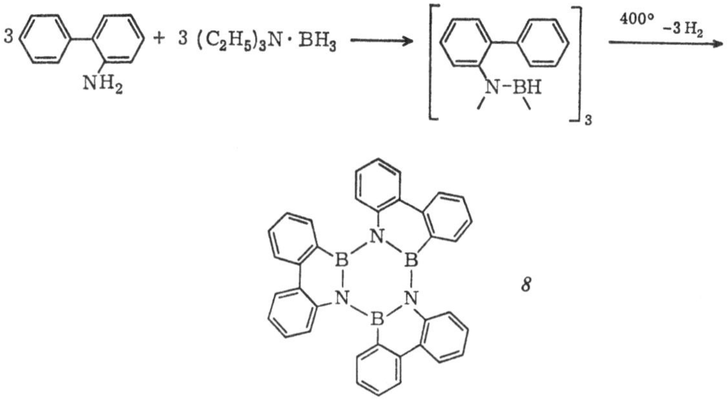

(26)

A borazine derivative has been obtained from the reaction of pyrazabole with o-phenylenediamine [Eq. (27)]. The resultant product is characterized by high stability [112].

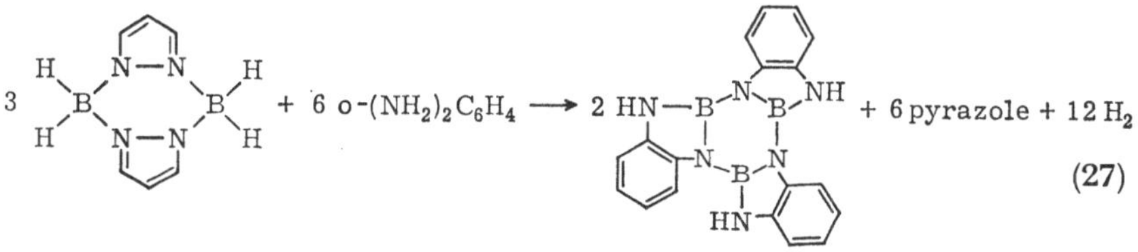

(27)

A most interesting borazaro derivative was prepared by the following reaction sequence [113] [Eq. (28)].

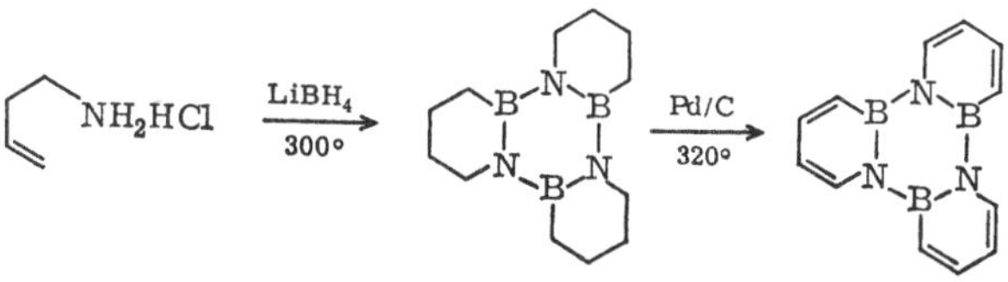

(28)

Other borazine reactions which have been described recently include the Grignard methylation of 1,3,5-trimethyl-2,4,6-trichloroborazine[114], the interaction of borazine and methylborazine with water [115], and ring cleavage of borazine by aniline [116]. The latter reaction yields N,N'-diphenyltriaminoborane by nucleophilic attack on the B-atoms:

$$[HNBH]_3 + 6 \, C_6H_5NH_2 \rightarrow 3 \, (C_6H_5NH)_2BNH_2 + 3 \, H_2 \qquad (29)$$

3. Substitution Reactions on Borazines

3.1. Substitution Reactions at the Nitrogen Atoms

Substituents on the nitrogen atoms of borazines include alkyl and aryl groups as well as hydrogen, lithium, chlorine, silicon and, in polyborazines, boron [4]; $>N-Na$ [117] and $>N-Mg$ bonds (cf. Section II.2.2.) have been postulated as reaction intermediates. Of these, H, Li, Cl and perhaps Si can be expected to be reactive towards substitution under conditions which will preserve the borazine structure. Substitution reactions on borazine nitrogen atoms carrying chlorine or silicon have not yet been described. Reactions on N—H groups of borazines have been reported to yield B—N bonds either directly under elimination of HX (X = H, halogen, NH$_2$, OSiR$_3$, etc., which are removed from a boron atom of another boron-containing molecule) or else after the hydrogen has first been replaced by a metal atom. Attempts to prepare other derivatives have been unsuccessful [28].

3.2. Substitution on the Boron Atom

In contrast to the scarcity of reports about N-substitution, B-substitution in borazines shows a nearly unlimited field of application. The possibilities of B-substitution have already been discussed in several reviews [1-4] and only some more recent developments will be considered here. Nearly all substitution reactions on the B-atoms of borazine may result in complete or partial substitution, depending on the molar ratios of the reactants.

3.2.1. Leading to the B-Haloborazines

While symmetrical B-chloro and B-bromoborazines result from the reactions of amines or amine-hydrohalides with the corresponding trihaloboranes, unsymmetrical representatives of these compounds have been obtained recently by reacting metal salts with B-trihydroborazines. Mixtures of mono- and di-haloborazine compounds have been obtained from borazine and HgCl$_2$ [118], and N-trialkylborazines and SnCl$_4$ or SnBr$_4$ respectively [119]. Unsymmetrical B-haloborazines have also been prepared from N-trialkylborazines and COCl$_2$, SO$_2$Cl$_2$, SOCl$_2$ and PCl$_3$ [120]. As the difference in reactivity of the B—H and B-halogen grouping is rather small with respect to most reactants, these unsymmetrical borazines cannot be considered to be mono- or difunctional like B-alkyl-B-haloborazines. Even B-fluoro- [31,107,121] and B-iodoborazine derivatives [30,46] have been prepared by direct ring synthesis, but normally such borazines are obtained most advantageously by transhalogenation reactions from the corresponding B-chloroborazines (preferably for the B—F compounds) or by iodination of the B—H borazines. The preparation of B-fluoroborazines from B-chloroborazines by TiF$_4$ [122] or SbF$_3$ [28] has been accomplished without utilization of a solvent; also, fluorination with NaF in acetonitrile [123] has been described. By these methods a variety of B-fluorinated borazines have been prepared [22,72–74]. Unsymmetrical B-butyl-B-fluoroborazines can be prepared accordingly, whereas, due to the exchange reactions of the B-methyl-groups, unsymmetrical B-methyl-B-fluoroborazines could be obtained only by partial methylation at the boron sites of fluoroborazines with methyllithium [124].

Iodination of B-hydridoborazines has given access to B-triiodoborazine compounds [125,126] as well as mono- and diiodoborazines [125] [Eq. (30)].

$$[RNBH]_3 + 3\,X_2 \rightarrow [RNBX]_3 + 3\,HX \tag{30}$$

This latter reaction was also applied to the preparation of B-tribromoborazines by direct bromination of borazine compounds [127].

3.2.2. Leading to B-Pseudohaloborazines

B-Cyano-, B-isocyanato- and B-isothiocyanato-borazines are preferably prepared by reacting the B-chloroborazines with silver cyanide [128], cyanate, or thiocyanate [22,72,129]. Also, B-azidoborazines have been obtained recently. The isolation of [HNBN]$_3$ has been claimed in a patent [130]. This work could not be reproduced due to the explosiveness of the compound [27,123]. However, several N-organosubstituted-B-azidoborazines have been made by reaction of symmetrical and unsymmetrical B-chloroborazines with NaN$_3$ in acetonitrile [22,123]. N-Organo-B-azidoborazines can be distilled in high vacuum if suitable precautions are taken [22].

3.2.3. Leading to Borazines with Unsaturated Side Chains

Interest in boron compounds with unsaturated side chains rests on the possible utilization of such compounds for further reactions in the side chains. Several of such borazines have been known [4], but new compounds of this type have been developed more recently.

B-Tris(ethinyl)borazines could be obtained by reacting B-halo-borazines and the Grignard compounds of acetylene derivatives [131,132] [Eq. (31)].

$$[RNBCl]_3 + 3\ R'C\equiv CMgBr \rightarrow [RNBC\equiv CR']_3 + 3\ BrMgCl \qquad (31)$$

Also, unsymmetrical substituted (mono- and di)ethinylborazines were described [133]. By catalytic hydrogenation, B-triethinylborazines have been converted to 2,4,6-tris(cis-methylvinyl)- and 2,4,6-tris(cis-phenyl-vinyl)borazines. The latter cannot be prepared by direct vinyl-Grignard reaction [134]. Alkenylborazines have also been obtained from the inter-action of vinyldichloroborane with amines [135], though this procedure does not seem to be a useful synthetic procedure since $CH_2=CHBCl_2$ is very sensitive towards oxygen. Pentamethylvinylborazine has been made from vinylmagnesium bromide and pentamethyl-B-monochloro-borazine [136]; fluorovinylborazines have been cited elsewere (see Section II.2.4). B-Triallylborazine was synthesized by a Grignard procedure [137].

Borazine compounds having exocyclic C=N double bonds can be obtained by reacting N-trialkylborazines with isonitriles [138] [Eq. (32)]

$$[RNBH]_3 + 3\ R'N\equiv C \rightarrow [RNB-CH=NR']_3 \qquad (32)$$

This reaction is also suitable for preparing unsymmetrical derivatives.

3.2.4. Leading to B-Aminoborazines

B-Aminoborazines are of particular interest for fundamental studies. In these compounds, boron is bonded to three nitrogen atoms with two different types of environment. B-Aminoborazines are also useful pre-cursors for the synthesis of thermally stable polymers. Quite a few poly-condensates of aminoborazines and copolymerisates with organic di-functional molecules have been described [4]. Of major interest are di-functional borazines yielding linear polycondensates. The condensation of 1,3,5-tris(2,6-dimethylphenyl)-2,4-dichloroborazine (cf. Section II.2.5) with aliphatic, aromatic, and heterocyclic diamines, as well as the pre-paration of the same linear polyborazines by transamination of 1,3,5-tris(2,6-dimethylphenyl)2,4-bis(diethyl-amino)borazine with diamines was studied [139].

$$n \quad \begin{array}{c} R \\ | \\ N \\ X-B \diagup \diagdown B-X \\ | \quad\quad | \\ R-N \diagdown \diagup N-R \\ B \\ | \\ H \end{array} \quad + n\, H_2N(CH_2)_xNH_2 + 2\, n\, (C_2H_5)_3N \rightarrow$$

$$\rightarrow \left[\begin{array}{c} R \\ | \\ N \\ -B \diagup \diagdown B-NH(CH_2)_xNH \\ | \quad\quad | \\ R-N \diagdown \diagup N-R \\ B \\ | \\ H \end{array} \right]_n \qquad (33)$$

By reactions of 1,2,3,5-tetramethyl-4,6-di(alkylamino)borazines with several diamines and 4,4'-diisocyanatodiphenylmethane, it was shown that the thermal stability of the bis(alkylamino)borazine depends on the substituents at the exocyclic nitrogens. Reactions with diamines led to polycondensates such as the ones cited above. However, reaction with diisocyanates gave different results depending on the substitution pattern of the exocyclic amino groups [140,141].

$$\begin{array}{c} CH_3 \\ | \\ N \\ HN-B \diagup \diagdown B-NHR \\ | \quad\quad | \\ CH_3-N \diagdown \diagup N-CH_3 \\ B \\ | \\ CH_3 \end{array} \quad + n\, OCN(X)NCO$$

(34)

$$\xrightarrow[25°]{R=C_6H_5} \quad C_6H_5NHB \begin{array}{c} CH_3\ O \\ | \quad \| \\ N-C \\ \diagup \quad\quad \diagdown N-X- \\ N-C \\ | \quad \| \\ CH_3\ O \end{array} \Big\rangle 100°\ \text{polycond.}$$

$$\xrightarrow{R=C_4H_9} \left[\begin{array}{c} CH_3 \\ | \\ N \\ -N-B \diagup \diagdown B-N-CONH(X)NHCO \\ | \quad\quad | \quad\quad | \\ C_4H_9\ CH_3N \diagdown \diagup NCH_3\ C_4H_9 \\ B \\ | \\ CH_3 \end{array} \right]_n$$

2-Bora-s-triazine compounds were formed from the anilinoborazines, whereas butylaminoborazines yield borazine-ureido polymers. The same condensation reactions have been studied starting from 1,3,5-triphenyl-2-

methylborazines [142]. Extension of the exocyclic amino groups of amino-borazines by reaction with diborane resulted in the formation of amino-borazine-boranes; however, the latter decompose readily by cleavage of the exocyclic borazine-nitrogen bonds [143] [Eq. (35)].

$$
n \;
\begin{array}{c}
R \\
| \\
\diagup N \diagdown \\
R'B \qquad B-NH_2BH_3 \\
| \qquad\qquad | \\
R-N \diagdown\;\diagup N-R \\
B \\
| \\
R'
\end{array}
\;\rightarrow n\;
\begin{array}{c}
R \\
| \\
\diagup N \diagdown \\
R'B \qquad B-H \\
| \qquad\quad | \\
R-N \diagdown\;\diagup N-R \\
B \\
| \\
R'
\end{array}
\;+ (HNBH)_n + n\,H_2 \qquad (35)
$$

Decomposition products of this latter reaction are the corresponding borazines, and polymeric or cyclic dimeric and trimeric B—N compounds, depending on the side chain substitution pattern.

Unsymmetrically substituted aminoborazines resulted from the reaction of amines with B-alkylthioborazines [144]. The first optically active borazine has been obtained from the reaction of 2,4,6-trichloroborazine and (+)2-butylamine [145] using triethylamine as a hydrogen chloride acceptor. B-Monoaminoborazine, $H_3N_3B_3H_2(NH_2)$ and isotopically labeled derivatives thereof have been prepared by photochemical reaction from borazine and ammonia in a gas phasereaction [146,147].

4. Miscellaneous Borazine Derivatives and Reactions

Photochemical reactions were employed to prepare diborazinylether [146] 9 and B-monoalkoxyborazines [148] from gaseous borazine/water and borazine/alcohol mixtures by u. v. irradiation.

$$
\begin{array}{c}
H\;\;H \qquad\qquad\qquad H\;\;H \\
\diagup B-N \diagdown \qquad\qquad \diagup N-B \diagdown \\
HN \qquad\quad B-O-B \qquad\quad NH \\
\diagdown B-N \diagup \qquad\qquad \diagdown N-B \diagup \\
H\;\;H \qquad\qquad\qquad H\;\;H
\end{array}
$$

9

B-Monohydroxyborazine [146] has not been isolated but was identified by spectroscopic methods. Borazine forms adducts with 3 moles of anhydrous hydrogenchloride and hydrogenbromide respectively:

$$\text{[structure]} + 3\,HX \rightarrow \text{[structure]} \tag{36}$$

Pyrolysis of these trimeric aminoboranes generally leads to dissociation into the starting materials. Some B-mono and B-dichloroborazine was also identified amongst the reaction products; also, polymeric materials were observed which resulted from B-haloborazine condensation. B-trihaloborazines were not obtained. Analogous reactions were observed with borazanaphtalene, $B_5N_5H_8$ [149] *10*, which forms an adduct with 5 HX.

10

B-Siloxyborazine derivatives are formed by reaction of B-mercapto-alkylborazines with $NaOSi(CH_3)_3$ [144] and their thermal stability has been investigated [150]. It was found that borazines with electron-donating groups on boron show reduced thermal stability as compared to hexaorganoborazines.

A uniquely substituted borazine is 1,3,5-trimethyl-2,4,6-tris(n-butyl-o-carboranyl)borazine which is obtained on reacting the 1,3,5-trimethyl-2,4,6-trichloroborazine with C-monolithio-C'-n-butyl-o-carborane [151] [Eq. (37)].

$$[CH_3NBCl]_3 + 3\ o\text{-}B_{10}H_{10}C_2(n\text{-}C_4H_9)Li \rightarrow [CH_3NB\text{-}o\text{-}B_{10}H_{10}C_2\text{-}n\text{-}C_4H_9]_3 \tag{37}$$

Interaction of anhydrous acetic acid and 1,3,5-triphenyl-2,4,6-trichloroborazine results in a quantitative yield of phenylacetamide. However, monofunctional B-acyl derivatives were obtained from penta-alkylmonochloroborazines and silver acetate or silver trifluoroacetate [152] in anhydrous ether [Eq. (38)].

$$R_3N_3B_3(C_5H_5)_3 + 3 \text{ (maleic anhydride)} \longrightarrow \text{[product]} \qquad (39)$$

Products resulting from a Diels-Alder reaction and from hydrostannation of B-cyclopentadienylborazines were described [152]. Monomeric Diels-Alder products were obtained from the reaction with maleic anhydride [Eq. (39)], polymeric products resulted from the interaction with chinone.

In general, B-alkoxyborazines are readily obtained from the interaction of the corresponding B-chloroborazines with sodium alkoxides. This method has been applied for the synthesis of symmetrical and unsymmetrical tetrafluoropropoxy- and propoxyborazines [153]. However, t-butanol may also be reacted directly [Eq. (40)].

$$[RNBCl]_3 + 3 \text{ t-}C_4H_9OH \rightarrow [RNBO\text{-t-}C_4H_9]_3 + 3 \text{ HCl} \qquad (40)$$

Polymers can be prepared by refluxing t-butylalkoxyborazines in the presence of boric acid [154]. Borazine-containing polymers have also been made by the following procedure [Eq. (41)] which is described in a patent [155].

$$[RNBX]_3 + 3 \text{ (R'O)}_2HPO \rightarrow [RNBOP(OR')_2]_3 \xrightarrow{\text{pyrol.}} \text{polymeric material} \qquad (41)$$

Decamethyl-B-B-biborazinyl is obtained by condensation of pentamethylchloroborazine with Na/K in inert solvents; direct bromination splits the B—B bond and yields pure pentamethylbromoborazine; this latter compound is not readily obtained by other methods, since B-methyl exchange reactions occur on distillation of B-methyl-B-halo-borazines [133].

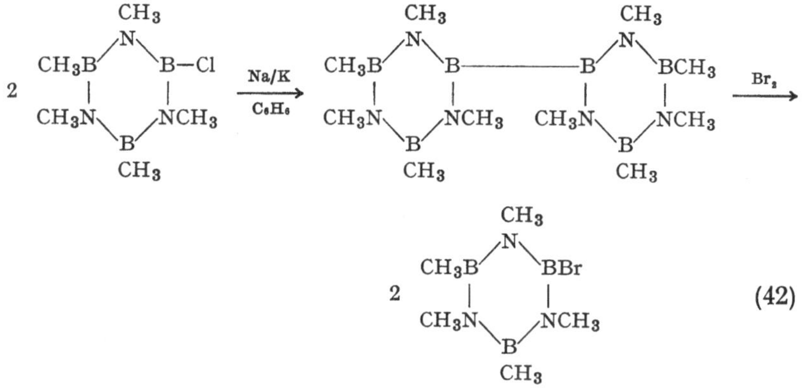

$$(42)$$

It has been claimed that B-B-polyborazines are formed in the pyrolysis of N-triphenyl-B-trimethylborazine [156]. B-Germylborazines have been prepared recently by reacting triphenylgermylpotassium, $(C_6H_5)_3GeK$, and germylpotassium with symmetrically and unsymmetrically substituted B-chloroborazines [157].

5. Borazine Complexes

One of the most fascinating developments in borazine chemistry has been the recent isolation of hexaalkylborazine chromiumtricarbonyl complexes. A reaction was observed when hexamethylborazine was reacted $Mo(CO)_6$ but no definite products could be isolated [158]. However, with extremely pure starting materials using $(CO)_3Cr(CH_3CN)_3$ or $(CO)_3(NH_3)_3Cr$ as reactants, the preparation of a borazine-metal complex has been effected [159,160]. Solvents other than dioxane did not lead to the desired product, and it has been suggested that the reaction proceeds via a dioxane-chromium tricarbonyl complex. The reaction appears to proceed by altering the existing equilibria, indicated in the reaction as depicted in Eq. (43).

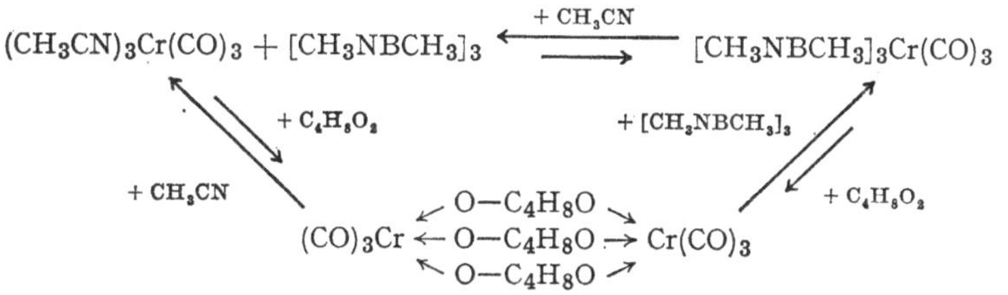

$$(43)$$

Borazine-chromiumtricarbonyl complexes derived from hexaalkyl-borazines with other alkyl substituents are obtained in similar manner. Analogous complexes containing B-chloroborazines could not be obtained. On the basis of dipole measurements, i.r. and n.m.r. data (which indicate a startling similarity between hexaalkylbenzene- and hexaalkyl-borazine-chromiumtricarbonyls) it seems reasonably well established that these compounds are π-complexes. It is worth noting that on reacting tris(dimethylamino)borane and $(CH_3CN)_3W(CO)_3$ in dioxane, tris (dimethylamino)borane-tricarbonyl-tungsten is obtained [161]. This latter material was also formulated as a π-complex.

A complex between hexamethylborazine and three molecules of nitroalkyl in which the NO_2-groups were resistant to catalytic reduction with Raney-Ni was described [162]. It was also reported that hexamethyl-borazine forms a 1:1 adduct with diglyme [163]. Another 1:1 complex was formed by reaction of 2,4,6-triphenylborazine with potassium in liquid ammonia and has been formulated as $[HNBC_6H_5]_3KNH_2$ [164].

B-Chloroborazines were found to give 1:2 and 1:4 complexes with pyridine; analogous but less stable compounds with α-picoline and 2,6-lutidine have also been reported [165]. A structure as depicted in *11* has been proposed for such materials.

11

6. s-Diazadiborines

The first s-diazadiborines were described in 1963 [166,167]. Since that time, several four-membered ring compounds of alternating B + N atoms with three-coordinated boron have been reported to result from a variety of different condensation, decomposition or rearrangement reactions of B—N derivatives. With two exceptions, these s-diazadiborine compounds are substituted by bulky substituents, and it has been shown [168] that steric hindrance is less significant in four- and eight-membered B—N rings than in the six-membered borazines.

Most of the s-diazadiborines described are not well characterized, some being liquids. It was claimed that 1,3-diphenyl-2,4-bis(triethyl-carbinyl)s-diazadiborine *12* was formed by pyrolysis of the products of the reaction between phenylisonitrile and triethylborane [169].

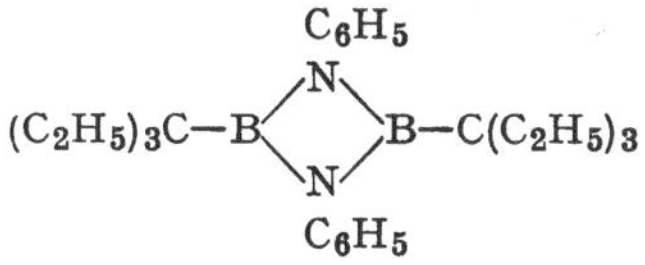

$$\begin{array}{c} C_6H_5 \\ \mid \\ (C_2H_5)_3C-B \underset{\underset{\displaystyle C_6H_5}{N}}{\overset{N}{\diamond}} B-C(C_2H_5)_3 \end{array}$$

12

However, it was reported later by the same authors that this structure was incorrect. The thermodynamically most stable final product, indeed, is a 5-membered B—N—C ring *13* [170].

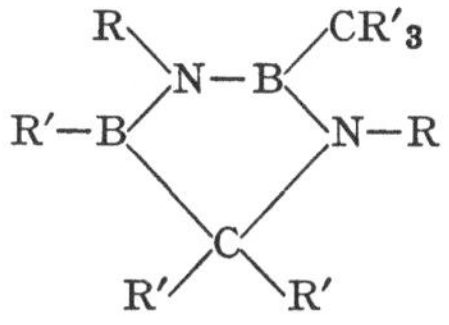

$$\begin{array}{c} R \diagdown \quad \diagup CR'_3 \\ N-B \\ R'-B \diagdown \quad \diagdown N-R \qquad R = aryl \\ \diagdown \quad \diagup \qquad\qquad R' = alkyl \\ C \\ R' \diagup \diagdown R' \end{array}$$

13

This observation promted a denial of the existence of s-diazadiborines [171], though an X-ray study had confirmed the four-membered ring structure in one of the compounds [172].

Methods for laboratory-scale production of s-diazadiborines have not yet been developed. Azidoboranes were reported to give good yields of s-diazadiborine compounds either by controlled photolytic decomposition in inert solvents or by controlled thermal decomposition without solvents [54,173].

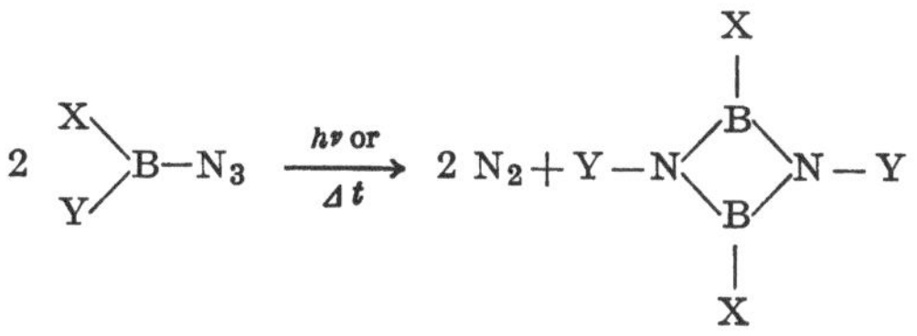

$$2 \;\; \overset{X}{\underset{Y}{\diagdown}} B-N_3 \xrightarrow{\; h\nu \text{ or} \;}_{\varDelta t} 2\,N_2 + Y-N \overset{\overset{\displaystyle X}{\mid}}{\underset{\underset{\displaystyle X}{\mid}}{\overset{B}{\underset{B}{\diamond}}}} N-Y \qquad (44)$$

$X = Y = N(CH_3)_2$ or $N(C_2H_5)_2$ or o-$CH_3C_6H_4$ or C_6H_5
$X = CH_3 \quad Y = C_6H_5$

Thermal decomposition of bis(dimethylamino)azidoborane in cyclohexene leads to the following compound [54] *14* [Eq. (45)].

$$2\,[(CH_3)_2N]_2BN_3 + 2\,C_6H_{10} \rightarrow 2\,N_2 + 2\,(CH_3)_2NH +$$

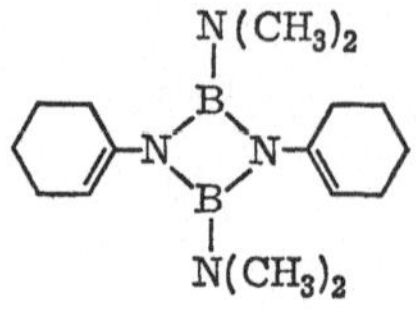

(45)

14

Possible mechanisms for the decomposition have been discussed [54]. Useful models for the reaction path were presented, but further studies will be needed in order to interpret unusual results in several instances. It should be noted, that tetraorgano-s-diazadiborines are sensitive towards oxygen.

The best characterized s-diazadiborine compound is 1,3-bis(trimethylsilyl)-2,4-bis(bis(trimethylsilyl)amino)-1,3-diaza-2,4-diborine. This compound was obtained from following reactions [105,106,108,174] [Eq. (46)].

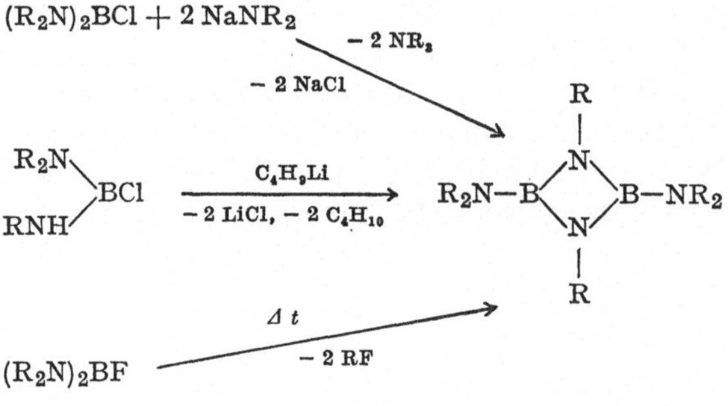

(46)

The structure of this compound has been confirmed by X-ray analysis [172]. The first s-diazadiborines were obtained by pyrolysis of tris(t-butyl-amino)borane [166,167]. Other less hindered tris(amino)boranes form aminoborazines or poly(aminoborazines); the nature of the product depends on pyrolysis time and utilized temperature [175]. Exocyclic B—N bonding appears to contribute to the stability of 1,3-diamino-diazaborines thereby explaining their chemical stability as compared to that of the corresponding tetraorganocompounds.

Representatives of the latter type were reputedly formed by thermal rearrangement of cyclic bis(iminoborane) type compounds with allyl substituents [176] [Eq. (47)].

$$Y = C_3H_5$$
$$R = CH_3 \text{ or } C_6H_5 \text{ or } CH=CH_2$$

(47)

Some hydrolysis and oxidation reactions of s-diazadiborines have been described [159], but no further data are available. This lack of information seems to confirm the view that the stability of the s-diazadiborine ring systems rests primarily on steric effects. While substitution with bulky substituents — even assuming that the B—N four-membered ring is stable in the transition state — presents a reaction which is not readily performed due to steric hindrance, substitution by smaller groups will not maintain the ring size. Unequivocal preparative proof will not be possible until more effective syntheses are developed for these compounds.

7. s-Tetraza-tetraborines

In general, thermal or base dehydrohalogenation of aliphatic amine adducts with trihaloboranes results in the formation of B-trihaloborazines (cf. Section II.2.2). In the case of tert.-butylamine (and some other amines with bulky organic groups) thermal dehydrohalogenation stops at the (alkylamino)dichloroborane stage. Subsequent dehydrohalogenation with bases yields a variety of products, amongst them s-tetraza-tetraborines. These substances were discovered and characterized by Turner and Warne [51,168,177,178]. It is essential for their preparation to start from the trihaloborane-triethylamine complexes, rather than to dehydrohalogenate any intermediate of the tert.-butylamine-trihaloborane system. Also, sufficient time of reflux must be maintained in order to complete the reaction. A detailed preparative procedure [Eq. (48)] is described in Inorganic Syntheses [179]:

$$4 \, X_3B{:}N(C_2H_5)_3 + 4 \, N(C_2H_5)_3 + 4 \, t{\cdot}C_4H_9NH_2 \xrightarrow{\text{toluene}}$$

$$\begin{array}{c} \text{structure} \end{array} \qquad + 8 \, (C_2H_5)_3NHX \qquad (48)$$

t·C₄H₉ diagram with ring, X = Cl or Br

Preparation of s-tetrazatetraborines by base-dehydrohalogenation of the BCl_3-t·C_4H_9 system at any intermediate level (i. e. (t·$C_4H_9NH_3$)⁺ BCl_4^-, t·$C_4H_9NH_2$·BCl_3 or t·$C_4H_9NHBCl_2$) has been described [46].

As noted above, (cf. II. 2.5) a s-tetrazatetraborine, (C_6H_5BN—i· C_4H_9)₄, has been reported to be formed along with the corresponding borazine derivative when the isobutylamine-phenyldichloroborane adduct was treated with triethylamine [49].

Neither catalytic reduction nor reduction by hydrides or alkylation (by RMgX, RLi or CH_2N_2) could be achieved: B-Hydrido-s-tetrazatetraborines are unknown. Also, transhalogenation to yield B-fluoro-derivatives could not be effected, but substitution of boron-bonded chlorine or bromine by pseudohalide groups (N_3, NCO, NCS or NCSe but not CN) proceeds smoothly and yields the tetrapseudohalo-substituted ring in nearly quantitative yield [51,180]. Since N-tert·butyl-B-hydrido- and B-fluoroborazine have been prepared (cf. II.2.5), this observation seems to indicate that substitution can be achieved only if a stable s-tetrazatetraborine derivative is formed. Substitutions that would require a change in the ring size in order to yield stable compounds will not occur. Further reactions of the pseudohalogeno group have yielded B-ureido-tetrazatetraborines [27].

III. B—N Ring Systems Involving Annular Boron-Boron Bonds

Boron-nitrogen compounds with alternating boron and nitrogen atoms are readily formed, apparently due to the high bond energy of the B—N linkage. In contrast, the synthesis of rings containing B—B bonds has been accomplished only by reaction of diborane(4) derivatives, in which the B—B bond is already preformed. It is well known that B—B bonds are stabilized by substituents which can donate π-electrons into boron orbitals [181–183]. It is apparent that the stability of B—N ring structures involving B—B bonds will be greatly facilitated when exocyclic amino- or alkoxy groups at the B atoms stabilize the B—B bonding. Diborane(4) derivatives with amino substituents are readily available [183]. The relative ease of splitting of the B—B bonds calls for mild conditions of condensation in order to minimize side reactions. Using these principles, Nöth and Abeler [184,185] abtained several ring systems involving B—B bonds. Only one of them, the five-membered 2,5-diaza-1,3,4-triborine ring, was the result of a synthesis directed to the formation of this system, while others containing the B—B sequence in six- and seven-membered rings have been isolated as products of side reactions.

1. 2,5-Diaza-1,3,4-triborines

Reaction of 1,2-dichloro-1,2-bis(dimethylamino)diborane(4) with N-dilithio bis- or tris(amino)borane derivatives resulted in several derivatives of the system named in the heading, according to the following equation:

$$
\begin{array}{c}
R \\
\diagdown \\
N-Li \\
Y-B \\
\diagup \\
N-Li \\
R
\end{array}
+
\begin{array}{c}
NR'_2 \\
Cl-B \\
| \\
Cl-B \\
NR'_2
\end{array}
\rightarrow
\begin{array}{c}
R \\
| \\
N-B-NR'_2 \\
Y-B \\
| \\
N-B-NR'_2 \\
| \\
R
\end{array}
+ 2\,LiCl \quad (49)
$$

Substituents R and Y may be parts of other (i.e. the triazaboradecaline) ring systems [184,185]. The yields range from 20 and 30% due to side reactions which are discussed below (cf. Section III.3). Attempts to synthesize such materials by thermal condensations according to Eq. (50)

$$
\begin{array}{c}
R \\
| \\
NR_2 \quad HN-B-R \\
Y-B \\
NR_2 \quad HN-B-R \\
| \\
R
\end{array}
\text{ or }
\begin{array}{c}
R \\
| \\
NH \quad R_2N-B-R \\
Y-B \\
NH \quad R_2N-B-R \\
| \\
R
\end{array}
\quad (50)
$$

did not yield the desired compounds. Rather decomposition of the alkylamine substituted reactants occurred and resulted in the formation of more stable products, i.e. borazines.

Substitution of one of the B-dialkylamino groups by Cl can be achieved with BCl_3 or $SiCl_4$ but the monochloro derivatives are quite unstable.

$$-B\begin{array}{c} N-B-NR_2 \\ | \\ N-B-NR_2 \end{array} \xrightarrow[BCl_3]{SiCl_4 \text{ or}} -B\begin{array}{c} N-B-Z \\ | \\ N-B-Cl \end{array} \qquad Z = NR_2 \text{ or } Cl$$

The dichloro compound is even less stable. It was obtained as an impure material from the reaction with BCl_3. Attempts to alkylate boron-bonded chlorine with RLi or RMgX were unsuccessful, as were attempts to alkylate the (B)-NR_2 groups with aluminium alkyls.

2. 1,4-Diaza-2,3,5,6-tetraborines

If a condensation is attempted to yield the 2,5-diaza-1,3,4-triborine compounds by base-dehydrohalogenation with triethylamine, trimethylamine is eliminated (R' = CH_3) and a condensation proceeds under formation of a tricyclic B—N ring system containing a six-membered B—N ring with two B—B sequences [185]. This material *15* is a derivative of 1,4-diaza-2,3,5,6-tetraborine.

15

3. 1,3,5-Triaza-2,4,6,7-tetraborines

Side reactions of the lithiation which in part produces 2,5-diaza-1,3,4-tetraborine derivatives (cf. Section III.1) result in the formation of derivatives of the seven-membered 1,3,5-triaza-2,4,6,7-tetraborine ring [185]. This reaction probably proceeds via the steps depicted in Eq. (51).

$$
\begin{array}{c}
R \\
|
\end{array}
\quad
R'{-}B{\Big\langle}\!\!\begin{array}{c} N{-}Li \\ \\ N{-}Li \\ | \\ R \end{array}
\quad + \quad
\begin{array}{c} Cl{-}B{-}NR_2 \\ | \\ R_2N{-}B{-}Cl \end{array}
\quad + \quad
\begin{array}{c} R \\ | \\ Li{-}N \\ \\ Li{-}N \\ | \\ R \end{array}\!\!{\Big\rangle}B{-}R'
\quad\longrightarrow
$$

$$
\begin{array}{c}
R_2N{\Big\langle}\,B{-}B\,{\Big\rangle}NR_2 \\
R{-}N \qquad\qquad N{-}R \\
| \qquad\qquad\qquad | \\
R'{-}B \qquad\qquad B{-}R' \\
/ \qquad\qquad\qquad \backslash \\
Li{-}N{-}R \qquad\quad R{-}N{-}Li
\end{array}
\quad\xrightarrow{-\,Li_2NR}\quad
\begin{array}{c}
R_2N{\Big\langle}\,B{-}B\,{\Big\rangle}NR_2 \\
R{-}N \qquad\qquad N{-}R \\
\backslash \qquad\qquad\quad / \\
R'{-}B \qquad\quad B{-}R' \\
\backslash \qquad\quad / \\
N \\
| \\
R
\end{array}
\qquad (51)
$$

IV. B—N Ring Systems with Annular Nitrogen-Nitrogen Bonds

Nitrogen-nitrogen bonds are not usually formed in B—N ring syntheses. Procedures for the preparation of B—N rings with N—N units normally utilize molecules that already contain N—N bonds, i.e. hydrazine or azo derivatives. However, there is a remarkable exception: the formation of the cyclotetrazenoborane system proceeds through the reaction of a boronimide with an azide, whereby a B—N and a N—N bond are formed (cf. Section IV.2).

1. 1,3,4-Triaza-2,5-diborines

The formation of a system containing two boron and three nitrogen atoms in a five membered ring cannot be expected to proceed in high yield from a spontaneous reaction. Indeed, preparations of this ring system utilize the most versatile reaction type applicable to B—N compounds: the transamination. Transamination usually proceeds by replacement of organo-substituted amino groups by less volatile amines [186–189] in an equilibrium reaction.

1,3,4-Triaza-2,5-diborines have been isolated from the reaction of B,B-bis(dialkylamino)diborylamine with 1,2-dialkylhydrazines [190]. Also, transamination of 1,2,4,5-tetraza-3,6-diborine derivatives with alkylamine [191,204] have been described; the former method seems to be more suitable [192].

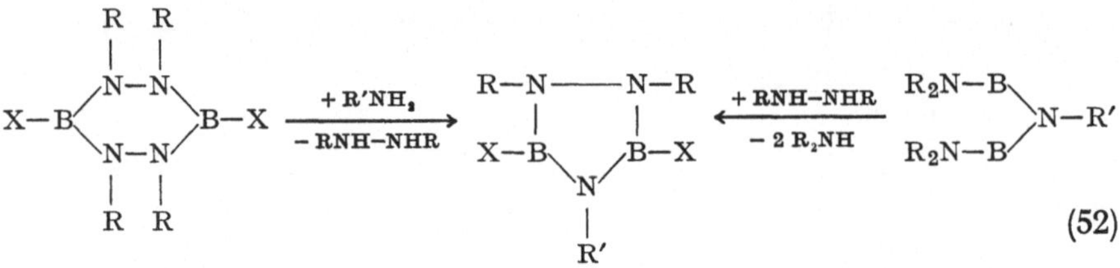

$$X-B \underset{\substack{N-N \\ \vert \quad \vert \\ R \quad R}}{\overset{\substack{R \quad R \\ \vert \quad \vert \\ N-N}}{\diagup\diagdown}} B-X \quad \xrightarrow[\text{$-$ RNH$-$NHR}]{\text{$+$ R'NH}_2} \quad$$

equation (52)

$$X = C_6H_5, \ R = CH_3, \ R' = \text{alkyl or H}$$

Metallation of 2,5-diphenyl-3,4-dimethyl-1,3,4-cyclotriaza-2,5-diborine (TADBH) has been accomplished with organolithium compounds. Subsequently the N-lithio compound (TADBLi) was reacted with metal salts and several interesting derivatives have been prepared [192] by this procedure. TADBLi not only reacts with diorganohaloboranes forming an exocyclic B—N bond, but also with $HgCl_2$ and $FeCl_2$ yielding the corresponding metal compounds $(TADB)_2M$ [Eq. (53)].

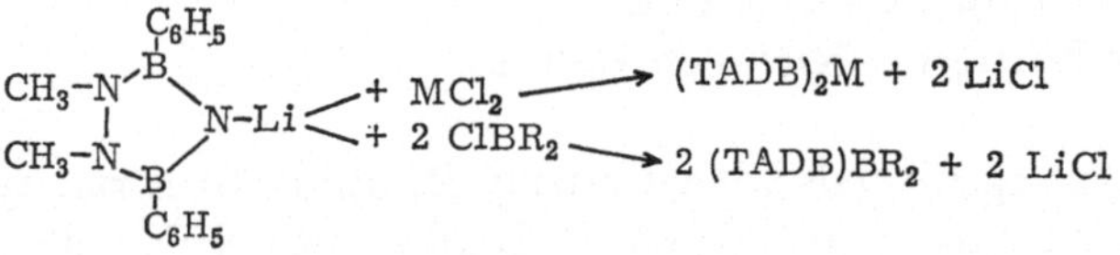

$$\begin{array}{c} CH_3-N\diagdown^{\overset{C_6H_5}{B}} \\ \qquad \qquad N-Li \\ CH_3-N\diagup_{\underset{C_6H_5}{B}} \end{array} \begin{array}{l} \xrightarrow{+ \ MCl_2} (TADB)_2M + 2 \ LiCl \\ \xrightarrow{+ \ 2 \ ClBR_2} 2 \ (TADB)BR_2 + 2 \ LiCl \end{array}$$

equation (53)

While the mercury compound shows localized N—Hg bonds, the TADB-system (which is isoelectronic with cyclopentadiene) forms a low-spin complex with iron. This was the first identified example of a sand-wich-compounds of a boron-nitrogen ring. The complex is prepared by adding an etheric solution of TADBLi to a suspension of $FeCl_2$ in tetrahydrofuran.

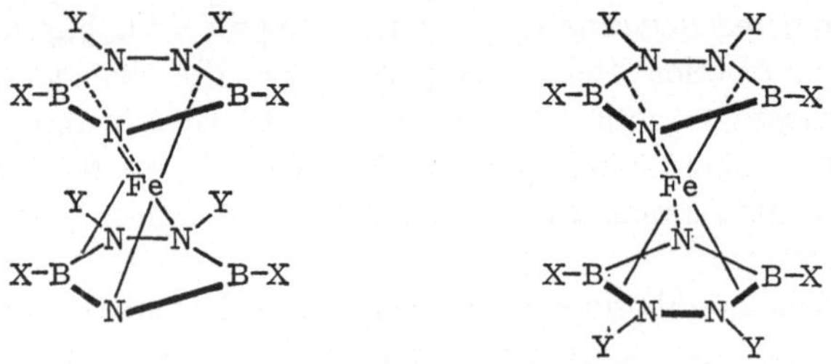

16   17

Possible structures of
bis(2,5-diphenyl-3,4-dimethyl-cyclo-1,3,4-triaza-2,5-diborinyl)-iron(II)

Bis(TADB)iron(II) is soluble in aromatic and etheric solvents and in CHCl₃. It is unsublimable, does not react with cyclopentadiene at 150°, and while it is relatively stable against bases, amines and alcohols, it is readily decomposed by acids.

2. Cyclotetrazenoboranes

This five-membered ring cenompassing four nitrogen and one boron atom is formed on reaction of an organic azide with a boronimide.

$$
\begin{array}{ccc}
\text{N}\equiv\text{N} & & \text{N}=\!=\!=\text{N} \\
\| & \longrightarrow & |\quad\quad| \\
\text{R}-\text{N}\quad\text{N}-\text{R}' & & \text{R}-\text{N}\diagdown\quad\diagup\text{N}-\text{R}' \\
\text{B} & & \text{B} \\
| & & | \\
\text{Y} & & \text{Y}
\end{array}
\tag{54}
$$

Ring closure results by formation of a B—N and N—N bond. The highly reactive boronimides, which supposedly are formed from many B—N compounds by elimination reactions, are unstable reaction intermediates. They are susceptible to side reactions or exchange reactions at the formally two-coordinated boron atoms. Therefore, reaction conditions will be significant if it is desired to obtain the pure compounds in high yield.

Boronimides are formed by thermal decomposition of primary amine-boranes and primary amine-trihaloboranes (cf. Section II) and also from organoazidoboranes [54]. Indeed, these sources have been shown to give high yields of cyclotetrazenoboranes [193-195]. When the boron compounds utilized in the formation of cycloterazenoborane by interaction with organic azides constitute cyclotriborazanes [196] or dialkylamine-boranes [196], resultant yields are smaller. This observation indicates that boronimides are not the major species in the decomposition sequence of these compounds. Also, yields are lower when the cyclotetrazenoborane results from a reaction of decaborane formula, with excess organoazide [193]. In cyclotetrazenoboranes, boron-bonded hydrogen atoms can be replaced by organic groups through Grignard-reactions [196] or by bromine with N-bromosuccinimide. An authoritative account of the chemistry of cycloterazenoboranes was presented by Morris and coworkers [7].

3. 1,2,4,5-Tetraza-3,6-diborines

Due to the symmetrical arrangement of the B- and N-atoms in this ring system it is readily synthesized [Eq. (55)]. The general route involves the reaction of a two-functional boron compound, RBX_2, with symmetrical or unsymmetrical hydrazine derivatives, which have at least one hydrogen atom at each nitrogen site [197,198]. Procedures are depicted in Eqs. (55) and (56).

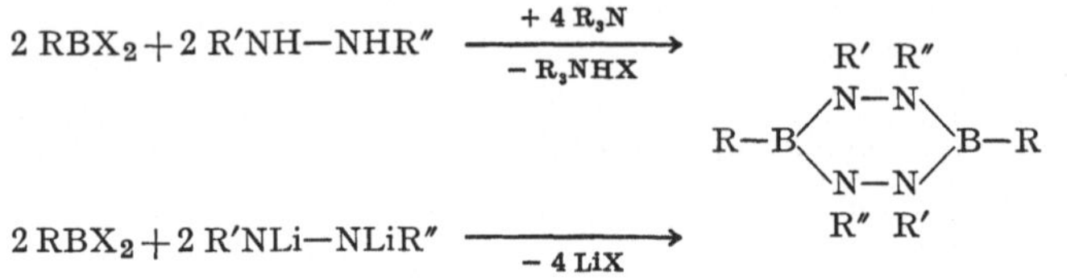

$$2\ RBX_2 + 2\ R'NH{-}NHR'' \xrightarrow[-\ R_3NHX]{+\ 4\ R_3N}$$

$$2\ RBX_2 + 2\ R'NLi{-}NLiR'' \xrightarrow[-\ 4\ LiX]{}$$

$$\tag{55}$$
$$\tag{56}$$

R = H, alkyl, aryl
X = H, NR$_2$, Cl
R', R'' = H alkyl, aryl

A unique synthetic method involves the reduction of azobenzene with diborane [197] [Eq. (57)]

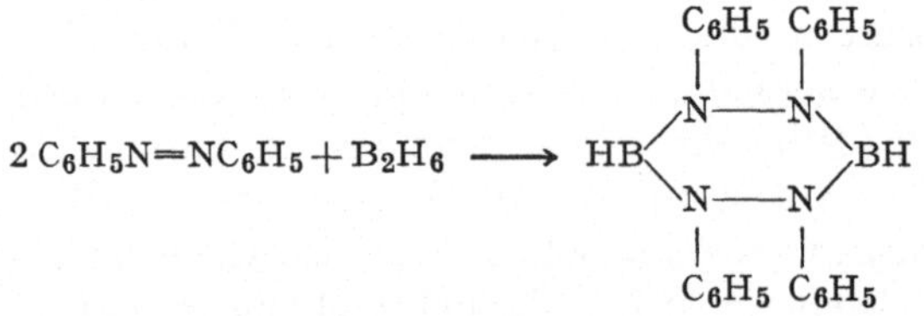

$$2\ C_6H_5N{=}NC_6H_5 + B_2H_6 \longrightarrow$$

$$\tag{57}$$

Tetraza-3,6-diphenyldiborine was obtained by hydrazinolysis of phenyl-bis(butylmercapto)borane [199]. However, the best laboratory method again appears to be a transamination of aminoboranes [197,200, 201] with hydrazine.

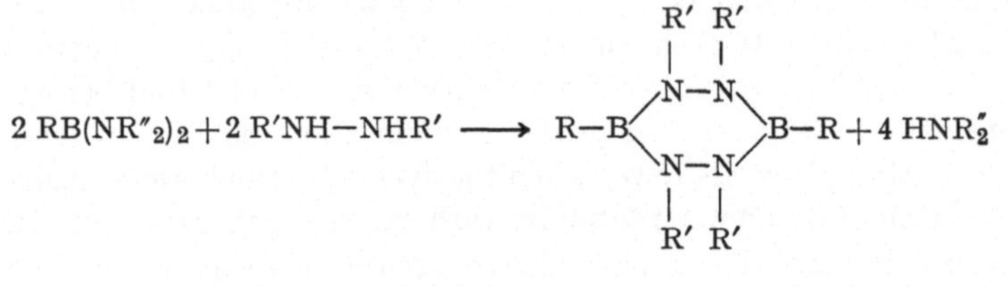

$$2\ RB(NR''_2)_2 + 2\ R'NH{-}NHR' \longrightarrow$$

$$+ 4\ HNR''_2 \tag{58}$$

This reaction has been the subject of a detailed study [201] and it was shown that optimal conditions and the reaction rates are depending on steric factors; yields range near 80%. On utilization of unsymmetrical hydrazines, one could expect the formation of isomers. However, the products of the reaction between bis(dimethylamino)organoboranes or dichloroorganoboranes with monoalkylhydrazines seem to be the 1,4-disubstituted derivatives. This result can be interpreted by steric and kinetic reasons [201]. Title compounds isolated so far are listed in Table 3.

Table 3. *1,2,4,5-Tetraza-3,6-diborines (position/substituent)*

N 1	N 2	B 3	N 4	N 5	B 6	m.p. °C	b.p./rnm.	Ref.
H	C_6H_5	H	H	C_6H_5	H	146 (dec.)		201)
H	H	C_6H_5	H	H	C_6H_5	150/253		197,199, 200)
C_6H_5	C_6H_5	H	C_6H_5	C_6H_5	H	112		197)
C_6H_5	H	C_6H_5	C_6H_5	H	C_6H_5	135		198)
C_6H_5	C_6H_5	C_6H_5	C_6H_5	C_6H_5	C_6H_5	156		197,198)
CH_3	H	CH_3	CH_3	H	CH_3		73/9	201)
CH_3	CH_3	CH_3	CH_3	CH_3	CH_3		52/9	191)
C_6H_5	H	CH_3	C_6H_5	H	CH_3	91—93		201)
CH_3	H	C_6H_5	CH_3	H	C_6H_5	87—91		201)
CH_3	CH_3	C_6H_5	CH_3	CH_3	C_6H_5	110		191)
CH_3	CH_3	CH_3	CH_3	CH_3	C_6H_5		79/10^{-3}	201)

Normally, hydrazinoborane derivatives with three-coordinated boron atoms are extremely sensitive towards moisture.

1,2,4,5-Tetraza-3,6-diborines can be considered as dimeric hydrazinoboranes; aggregation of hydrazinoboranes, $(-RB-NR'-NR'-)_n$, is not limited to $n = 2$, but may proceed further depending on substitution.

Recently, *Tetrameric Hydrazinoboranes* have been isolated

a) from the pyrolysis of hydrazine-tert. butylborane [202]

$$4 \text{ t} \cdot C_4H_9B \overset{H}{\underset{H}{\leftarrow}} NH_2NH_2 \xrightarrow{\Delta t} (\text{t} \cdot C_4H_9BNHNH)_4 + 8 H_2 \qquad (59)$$

and

b) by hydrazinolysis of bis(dimethylamino)borane with monomethylhydrazine or with 1,2-dimethylhydrazine [201]:

$$4 \text{ HB}[N(CH_3)_2]_2 + 4 \text{ CH}_3NH-NHR \rightarrow [HBNCH_3NR]_4 + 8 \text{ HN}(CH_3)_2 \qquad (60)$$

R = H or CH_3

Tetrameric hydrazinoboranes exhibit a polyhedral cage structure involving tetra-coordinated boron (structure *18*) which was confirmed by X-ray studies [203].

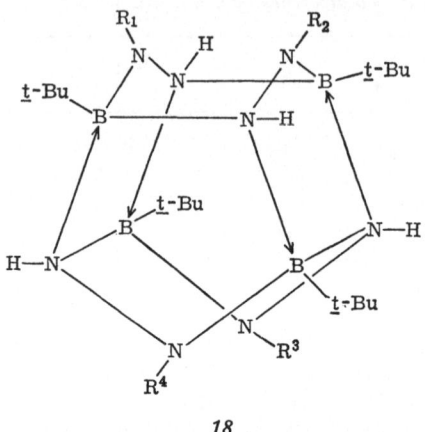

18

Dimeric 1,2,4,5-tetraaza-3,6-bis(tert.butyl)-diborine

The nitrogen atoms are either trivalent or tetravalent. Hydrogen attached to the trivalent nitrogens reacts with methylisocyanate by insertion of CH_3NCO between boron and nitrogen on either 1,2,3 or all 4 trivalent nitrogen sites [202]. It should be noted that the higher aggregation due to tetracoordination results in a drastic change of the chemical properties of hydrazinoboranes [201,204]. 1,2,4,5-Tetraza-3,6-diborines react with hydrogen chloride yielding adducts with two (or in one case three) molar equivalents of HCl, or by splitting of the B—N bonds according to the basicity of the parent hydrazine [204] [Eq. (61)].

$$2[R''HN-NR'H_2]Cl + 2\,RBCl_2 \tag{61}$$

Tetrameric hydrazinoboranes give adducts with two molar equivalents of HCl forming salt-like structures.

While 1,2,4,5-tetraza-3,6-diborines give adducts with two moles of Lewis acids (BF_3, BCl_3, $FeCl_3$), tetrameric hydrazinoboranes were found to be unreactive, or else they add only molar equivalent of BCl_3, according to their substitution pattern [204]. Differences in reactivity of the latter compounds are attributed to sterical reasons. Nöth and Regnet [201] have stated, that these tetrameric compounds are the connecting links between dimeric and polymeric hydrazinoboranes. The degree of aggregation appears to depend on steric factors and on the Lewis acidity of the hydrazinoboranes. The two compounds $(HBNHNH)_n$ and $(CH_3BNHNH)_n$ have consequently been found to be polymeric [201].

4. 1,2,4,6-Tetraza-3,5,7-triborines

The only representative of this seven-membered ring system was obtained by a transamination reaction using a derivative of triazaboradecaline [Eq. (62)]. The boron atom and the two nitrogens of the triazaboradecaline (a system which is most suitable for further reactions, due to its stability) form part of the seven-membered ring [205] as depicted in structure 19.

$$(62)$$

$$19$$

V. References

[1] Sheldon, J. C., Smith, B. C.: Quart. Rev. (London) 14, 200 (1960).
[2] Mellon, E. K., Lagowski, J. J.: Advan. Inorg. Chem. Radiochem. 5, 259 (1963).
[3] Niedenzu, K., Dawson, J. W.: Boron-Nitrogen-Compounds. Berlin–Heidelberg–New York: Springer 1965.
[4] Steinberg, H., Brotherton, R. J.: Organoboron Chemistry, Vol. 2. Boron-Nitrogen and Boron-Phosphorus Compounds. New York: Interscience 1966.
[5] Geanangel, R. A., Shore, S. G.: In: Preparative Inorganic Reactions, Vol. 3, p. 123 (W. L. Jolly, ed.). New York: Interscience 1966.
[6] Niedenzu, K., Dawson, J. W.: In: The Chemistry of Boron and its Compounds, p. 443 (E. L. Muetterties, ed.). New York: Wiley 1967.
[7] Finch, A., Leach, J. B., Morris, J. H.: Organometal. Chem. Rev. A 4, 1 (1969).
[8] Paetzold, P. I., Simson, W. M.: Angew. Chem. 78, 825 (1966); Angew. Chem. Intern. Ed. Engl. 5, 842 (1966).

9) — Maisch, H.: Chem. Ber. *101*, 2870 (1968).
10) — Stohr, G.: Chem. Ber. *101*, 2874 (1968).
11) — Z. Anorg. Allgem. Chem. *326*, 64 (1963).
12) Wiberg, E.: Naturwissenschaften *35*, 182 (1948).
13) Stock, A., Pohland, E.: Ber. *59*, 2215 (1926).
14) Haworth, D. T., Hohnstedt, L. F.: Chem. Ind. (London) *1960*, 559.
15) Niedenzu, K., Dawson, J. W.: Inorg. Syn. *10*, 142 (1967).
16) Schaeffer, R., Steindler, M., Hohnstedt, L. F., Smith, S. H. Jr., Eddy, L. P., Schlesinger, H.: J. Am. Chem. Soc. *76*, 3303 (1954).
17) Bissot, T. C., Parry, R. W.: J. Am. Chem. Soc. *77*, 3481 (1955).
18) Beachley, O. T., Jr.: Inorg. Chem. *7*, 701 (1968).
19) — Inorg. Chem. *6*, 870 (1967).
20) Bonham, J., Drago, R. S.: Inorg. Syn. *9*, 8 (1967).
21) Niedenzu, K., Sawodny, W., Watanabe, H., Dawson, J. W., Totani, T., Weber, W.: Inorg. Chem. *6*, 1453 (1967).
22) Meller, A., Wechsberg, M.: Monatsh. Chem. *98*, 513 (1967).
23) Totani, T., Watanabe, H., Kubo, M.: Spectrochim. Acta *25A*, 585 (1969).
24) Jennings, J. R., Wade, K.: J. Chem. Soc. *1968*, 1946.
25) Horn, E. M., Lang, K.: Germ. Pat. 1.161.890 (1964).
26) Brown, M. P., Heseltine, R. W., Johnson, D. W.: J. Chem. Soc. *A 1967*, 597.
27) Meller, A.: unpublished results.
28) Laubengayer, A. W., Watterson, K., Bidinosti, D. R., Porter, R. F.: Inorg. Chem. *2*, 519 (1963).
29) Nöth, H.: Z. Naturforsch. *16b*, 618 (1961).
30) Meller, A., Marecek, H., Batka, H.: Monatsh. Chem. *98*, 2135 (1967).
31) Harris, J. J., Rudner, B.: Inorg. Chem. *8*, 1258 (1969).
32) Brown, C. A., Laubengayer, A. W.: J. Am. Chem. Soc. *77*, 3699 (1955).
33) Niedenzu, K., Dawson, J. W.: Inorg. Syn. *10*, 139 (1967).
34) Ryschkewitsch, G. E., Harris, J. J., Sisler, H. H.: J. Am. Chem. Soc. *80*, 4515 (1958).
35) Harris, J. J.: J. Org. Chem. *26*, 2155 (1961).
36) Meller, A., Schlegel, R.: Monatsh. Chem. *96*, 1209 (1965).
37) — Egger, H.: Monatsh. Chem. *97*, 790 (1967).
38) Wagner, R. I., Bradford, J. L.: Inorg. Chem. *1*, 99 (1962).
39) — — Inorg. Chem. *1*, 93 (1962).
40) — US. Pat. 3.255.244 (June 7th, 1966).
41) Boone, J. L., Willcockson, G. W.: Inorg. Chem. *5*, 311 (1966).
42) — US. Pat. 3.317.596 (1967).
43) Roubal, R. K.: Univ. Microfirms (Ann Arbor/Mich.) Order No. 65—11655 Dissertation Abstr. *26*, 2461 (1965).
44) Rothgery, E. F., Hohnstedt, L. F.: Inorg. Chem. *6*, 1065 (1967).
45) Butcher, I. M., Currell, B. R., Gerrard, W., Sharma, G. K.: J. Inorg. Nucl. Chem. *27*, 817 (1965).
46) — Gerrard, W.: J. Inorg. Nucl. Chem. *27*, 823 (1965).
47) Currell, B. R., Khodabocus, M.: J. Inorg. Nucl. Chem. *28*, 371 (1966).
48) Butcher, I. M., Currell, B. R., Sharma, G. K.: J. Inorg. Nucl. Chem. *28*, 2137 (1966).
49) Currell, B. R., Gerrard, W., Khodabocus M.: J. Organometal. Chem. (Amsterdam) *8*, 411 (1967).
50) Turner, H. S., Warne, R. J.: Chem. Ind. (London) *1958*, 526.
51) — — J. Chem. Soc. *1965*, 6421.
52) Paetzold, P. I.: Z. Anorg. Allgem. Chem. *326*, 47 (1963).

53) Meller, A.: Monatsh. Chem. 99, 1649 (1968).
54) Paetzold, P. I.: Fortschr. Chem. Forsch. 8, 437 (1967).
55) Niedenzu, K., Harrelson, D. H., Dawson, J. W.: Chem. Ber. 94, 671 (1961).
56) Meller, A., Batka, H.: unpublished results.
57) Wiberg, N., Raschig, F., Schmid, K. H.: J. Organometal. Chem. (Amsterdam) 10, 29 (1967).
58) Niedenzu, K., Fritz, P., Weber, W.: Z. Naturforsch. 22b, 225 (1967).
59) Haasnoot, J. G., Groeneveld, W. I.: Inorg. Nucl. Chem. Letters 3, 597 (1967).
60) Leeuwen, van, P. W. N. M., Groeneveld, W. L.: Rec. Trav. Chim. 86, 593 (1967).
61) Turner, H. S.: Chem. Ind. (London) 1958, 1405.
62) Lagowski, J. J., Thompson, P. G.: Proc. Chem. Soc. (London) 1959, 301.
63) Massey, A. G., Park, A. J.: J. Organometal. Chem. (Amsterdam) 2, 461 (1964).
64) Meller, A., Wechsberg, M., Gutmann, V.: Monatsh. Chem. 96, 388 (1965).
65) Leffler, A. J.: Inorg. Chem. 3, 145 (1964).
66) — US. Pat. 3.119.864 (1964).
67) Meller, A., Maringgele, W.: Monatsh. Chem. 99, 2504 (1968).
68) — Ossko, A.: Monatsh. Chem. 100, 1187 (1969).
69) Gridina, V. F., Klebanski, A. L., Bartashev, V. A.: Zh. Obshh. Khim. 34, 1401 (1964).
70) Nayar, V. S. V., Peacock, R. S.: Nature 207, 630 (1965).
71) Meller, A., Gutmann, V., Wechsberg. M.: Inorg. Nucl. Chem. Letters 1, 79 (1965).
72) — Wechsberg, M., Gutmann, V.: Monatsh. Chem. 97, 619 (1966).
73) Glemser, O., Elter, G.: Z. Naturforsch. 21b, 1132 (1966).
74) Meller, A., Wechsberg, M., Gutmann, V.: Monatsh. Chem. 97, 1163 (1966).
75) Klanica, A. J., Faust, J. P., King, C. S.: Inorg. Chem. 6, 840 (1967).
76) Lappert, M. F.: In: The Chemistry of Boron and its Compounds, p. 485 (E. L. Muetterties, ed.). New York: Wiley 1967.
77) Meller, A., Ossko, A.: Monatsh. Chem. 99, 1217 (1968).
78) — Marecek, H.: Monatsh. Chem. 99, 1355 (1968).
79) Brotherton, R. J., McCloskey, A. L.: Advan. Chem. Ser. 42, 131 (1964).
80) Lappert, M. F.: In: Developments in Inorganic Polymer Chemistry, p. 23 (M. F. Lappert and G. J. Leigh, ed.). Amsterdam: Elsevier 1962.
81) Meller, A., Schaschel, E.: Inorg. Nucl. Chem. Letters 2, 41 (1966).
82) Brown, M. P., Mellor, B. G., Silber, H. B.: Brit. Pat. 1.050.434 (1966).
83) Horn, E. M., Lang, K.: Germ. Pat. 1.156.071 (1963).
84) Grace, A., Powell, P.: J. Chem. Soc. A 1966, 673.
85) Currell, B. R., Gerrard, W., Khodabocus, M.: Chem. Commun. 1966, 77.
86) Burch, J. E., Gerrard, W., Mooney, E. F.: J. Chem. Soc. 1962, 2200.
87) Semlyen, J. A., Flory, P. J.: J. Chem. Soc. A 1966, 191.
88) — Philips, C. S. G.: J. Chromatog. 18, 1 (1965).
89) Bartlett, R. K., Turner, H. S., Warne, R. J., Young, M. A., Lawrenson, I. J.: J. Chem. Soc. A 1966, 479.
90) — — Brit. Pat. 1.009.362 (1965).
91) Nagasawa, K.: Inorg. Chem. 5, 442 (1966).
92) Nakagawa, T., Nagasawa, K.: Japan. Pat. 4.593 (1967) (C 214 E 4); C. A. 67, 32772 c.
93) — — Japan. Pat. 19.184 (1966) (C 216 E 4); C. A. 66, 38041 q.
94) — — Japan. Pat. 19.185 (1966) (C 216 E 4); C. A. 66, 38042 r.
95) Schlesinger, H. I., Ritter, D. M., Burg, A. B.: J. Am. Chem. Soc. 60, 1296 (1938).
96) Philips, C. S. G., Powell, P., Semlyen, J. A.: J. Chem. Soc. 1963, 1202.

97) Beachley, O. T., Jr.: Inorg. Chem. *8*, 981 (1969).
98) Philips, C. S. G., Powell, P., Semlyen, J. A., Timms, P. L.: Z. Anal. Chem. *197*, 202 (1963).
99) Schlesinger, H. I., Horvitz, L., Burg, A. B.: J. Am. Chem. Soc. *58*, 409 (1936).
100) Paetzold, P. I., Habereder, P. P., Müllbauer, R.: J. Organometal. Chem. (Amsterdam) *7*, 51 (1967).
101) Nakagawa, T., Watanabe, H., Nagasawa, K., Totani, T.: Japan. Pat. 30.170 and 30.171 (1964); C. A. *62*, 13179 g.
102) — — — — Japan. Pat. 7898 (1965); C. A. *63*, 13314 e.
103) Horn, E. M.: Germ. Pat. 1.201.838 (1965).
104) — Niederpruem, H.: Germ. Pat. 1.206.899 (1965).
105) Geymayer, P., Rochow, E. G.: Monatsh. Chem. *97*, 429 (1966).
106) — — Wannagat, U.: Angew. Chem. *76*, 499 (1964); Angew. Chem. Intern. Ed. Engl. *3*, 633 (1964).
107) Sujishi, S., Witz, S.: J. Am. Chem. Soc. *79*, 2447 (1957).
108) Russ, C. R., McDiarmid, A. G.: Angew. Chem. *76*, 500 (1964).
109) Niedenzu, K., Magin, G.: Z. Naturforsch. *20b*, 604 (1965).
110) Koester, R., Hatori, S., Morita, Y.: Angew. Chem. *77*, 719 (1965).
111) — Iwasaki, K., Hattori, S., Morita, Y.: Liebigs Ann. Chem. *720*, 23 (1968).
112) Trofimenko, S.: J. Am. Chem. Soc. *89*, 4948 (1967).
113) Culling, G. C., Dewar, M. J. S., Marr, P. A.: J. Am. Chem. Soc. *86*, 1125 (1964).
114) Fednewa, E. M., Kryukova, I. V., Alpatova, V. I.: Zh. Neorgan. Khim. *11*, 2058 (1966); C. A. *66*, 2609 c.
115) Weibrecht, W. E.: Univ. Microfilms (Ann Arbor/Mich.) Order No. 64—8762, Dissert. Abstr. *26*, 5028 (1966).
116) Kreutzberger, A., Ferris, F. C.: J. Org. Chem. *30*, 360 (1965).
117) Compton, R. D., Lagowski, J. J.: Inorg. Chem. *7*, 1234 (1968).
118) Maruca, R., Beachley, O. T. Jr., Laubengayer, A. W.: Inorg. Chem. *6*, 575 (1967).
119) Anderson, G. A., Lagowski, J. J.: Chem. Commun. *1966*, 649.
120) John, K.: Germ. Pat. 1.204.255 (1965) also French. Pat. 1.448.479 (1966).
121) Wiberg, E., Horeld, G.: Z. Naturforsch. *6b*, 338 (1951).
122) Niedenzu, K., Beyer, H., Jenne, K.: Chem. Ber. *96*, 2649 (1963).
123) Muszkat, K. A., Hill, L., Kirson, B.: Israel J. Chem. *1*, 27 (1963).
124) Meller, A., Wojnowska, M., Marecek, H.: Monatsh. Chem. *100*, 175 (1969).
125) Muszkat, K. A., Kirson, B.: Israel J. Chem. *1*, 150 (1963).
126) Meller, A., Wechsberg, M.: Monatsh. Chem. *98*, 690 (1967).
127) Riley, R. F., Schack, C. J.: Inorg. Chem. *3*, 1651 (1964).
128) Gutmann, V., Meller, A., Schaschel, E.: Monatsh. Chem. *95*, 1188 (1964).
129) Lappert, M. F., Pyszora, H.: J. Chem. Soc. *1963*, 1744.
130) Keith, J. N., Muszkat, St. F., Klein, M. J.: US. Pat. 3.394.999.
131) Watanabe, H., Totani, T., Yoshizaki, T.: Inorg. Chem. *4*, 657 (1965).
132) Nakagawa, T., Watanabe, H., Totani, T.: Japan. Pat. 21.778 (1966) (C 216 E 4); C. A. *66*, 46478 c (1967).
133) Meller, A., Marecek, H.: Monatsh. Chem. *99*, 1666 (1968).
134) Yoshizaki, T., Watanabe, H., Nagasawa, K., Totani, T., Nakagawa, T.: Inorg. Chem. *4*, 1016 (1965).
135) Braun, J., Normant, H.: Bull. Soc. Chim. France *1966*, 2557.
136) Klanica, A. J., Faust, J. P.: Inorg. Chem. *7*, 1037 (1968).
137) Fan Chieh Ts'eng, Li-chui Chiang, Hua Hsue Hsueh Pao: *31*, 111 (1965); C. A. *63*, 4317 g.
138) Meller, A., Batka, H.: will be published in Monatsh. Chem.

139) Turner, J. M.: J. Chem. Soc. A *1966*, 401, 410, 417.
140) Proux, Y., Clement, R.: Compt. Rend. Acad. Sci. Paris, Ser. C 264, 2123 (1967).
141) Clement, R., Proux, Y.: Bull. Soc. Chim. France *1969*, 558.
142) Korshak, V. V., Bekasova, N. I., Komarova, L. G.: Izv. Akad. Nauk. SSSR, Ser. Khim. 1965, 1462.
143) Meller, A.: Monatsh. Chem. *99*, 1670 (1968).
144) Mikhailov, B. M., Galkin, A. F.: Izv. Akad. Nauk. SSSR, Ser. Khim. *1968*, 407.
145) Haworth, D. T., Gould, G. F.: Chem. Ind. (London) 1967, 2113.
146) Lee, G. H., Porter, R. F.: Inorg. Chem. *6*, 648 (1967).
147) Porter, R. F., Young, E. S.: Inorg. Chem. *7*, 1306 (1968).
148) Nadler, M., Porter, R. F.: Inorg. Chem. *6*, 1739 (1967).
149) Laubengayer, A. W., Beachley, O. T., Porter, R. F.: Inorg. Chem. *4*, 578 (1965).
150) Vasilenko, N. A., Galkin, A. F., Mikhailov, B. M., Prowednikow, A. N.: Izv. Akad. Nauk. SSSR, Ser. Khim. *1968*, 2519.
151) Boone, J. L., Brotherton, R. J., Petterson, L. L.: Inorg. Chem. *4*, 910 (1965).
152) Gutmann, V., Meller, A.: Oesterr. Chemiker-Ztg. *66*, 324 (1965).
153) Meller, A., Schlegel, R., Gutmann, V.: Monatsh. Chem. *95*, 1564 (1964).
154) Turner, H. S., Larcombe, B. E.: Brit. Pat. 973.000 (1964).
155) Gerrard, W., Mooney, E. F.: Brit. Pat. 1.015.782 (1966).
156) Korshak, V. V., Zamyatina, V. A., Bekasova, N. I., Komarova, L. G.: Izv. Akad. Nauk. SSSR, Ser. Khim. *1964*, 2223.
157) Amberger, E., Stoeger, W.: J. Organometal. Chem. (Amsterdam) *17*, 287 (1969).
158) Fedneva, E. M., Kryukova, I. V.: Dokl. Akad. Nauk. SSSR *170*, 831 (1966).
159) Prinz, R., Werner, H.: Angew. Chem. 79, 63 (1967); Angew. Chem. Intern. Ed. Engl. *6*, 91 (1967).
160) Werner, H., Prinz, R., Deckelmann, E.: Chem. Ber. *102*, 59 (1969).
161) Schmid, G., Nöth, H., Deberitz, G.: Angew. Chem. *80*, 282 (1968).
162) Zamyatina, V. A., Oganesyan, R. M., Sevostyanova, V. V., Sidorov, T. A.: Izv. Akad. Nauk. SSSR, Ser. Khim. *1964*, 1881.
163) Fedneva, E. M., Kryukova, I. V.: Zh. Neorgan. Khim. *13*, 1467 (1968).
164) Compton, R. D., Koehl, H., Lagowski, J. J.: Inorg. Chem. *6*, 2265 (1967).
165) Lappert, M. F., Srivastava, G.: J. Chem. Soc. A *1967*, 602.
166) — Majumdar, M. K.: Proc. Chem. Soc. *1963*, 88.
167) — — Advan. Chem. Ser. *42*, 208 (1964).
168) Turner, H. S., Warne, R. J.: Advan. Chem. Ser. *42*, 290 (1964).
169) Casanova, J. Jr., Kieffer, H. R., Kuwada, D., Boultin, A. H.: Tetrahedron Letters *12*, 703 (1965).
170) — — J. Org. Chem. *34*, 2579 (1969).
171) Matteson, D. S.: Organometal. Chem. Rev. B *4*, 290 (1968).
172) Hess, H.: Angew. Chem. Intern. Ed. Engl. *6*, 975 (1967).
173) Paetzold, P. I.: Z. Anorg. Allgem. Chem. *326*, 53 (1963).
174) Wells, R. L., Collins, A. L.: Inorg. Chem. *7*, 419 (1968).
175) Aubrey, D. W., Lappert, M. F.: J. Chem. Soc. *1959*, 2927.
176) Bubnov, Y. N.: Zh. Obshch. Khim. *38*, 260 (1968).
177) Turner, H. S., Warne, R. J.: Proc. Chem. Soc. *1962*, 69.
178) — — Brit. Pat. 937.430 (1963).
179) Meller, A., Schaschel, E.: Inorg. Syn. *10*, 144 (1967).
180) Turner, H. S., Warne, R. J.: Brit. Pat. 946.989 (1964).
181) Nöth, H., Meister, W.: Chem. Ber. *95*, 515 (1961).
182) Brotherton, R. J., McCloskey, A. L., Petterson, L. L., Steinberg, H.: J. Am. Chem. Soc. *82*, 6242 (1960).

A. Meller

183) — In: Progress in Boron Chemistry, p. 1 (H. Steinberg and A. L. McCloskey, ed.). Oxford: Pergamon Press 1964.
184) Nöth, H., Abeler, G.: Angew. Chem. 77, 506 (1965); Angew. Chem. Intern. Ed. Engl. 4, 522 (1965).
185) — — Chem. Ber. 101, 969 (1968).
186) Niedenzu, K., Harrelson, D. H., Dawson, J. W.: Chem. Ber. 94, 671 (1961).
187) English, W. D., McCloskey, A. L., Steinberg, H.: J. Am. Chem. Soc. 83, 2122 (1961).
188) Aubrey, D. W., Lappert, M. F.: Proc. Chem. Soc. (London) 1960, 148.
189) Nöth, H.: Z. Naturforsch. 16b, 470 (1961).
190) Niedenzu, K., Jenne, H., Fritz, P. W.: Angew. Chem. 76, 535 (1964); Angew. Chem. Intern. Ed. Engl. 3, 514 (1964).
191) — Regnet, W.: Z. Naturforsch. 18b, 1138 (1963).
192) — — Z. Anorg. Allgem. Chem. 352, 1 (1967).
193) Greenwood, N. N., Morris, J. H.: J. Chem. Soc. 1965, 6205.
194) Morris, J. H., Perkins, P. G.: J. Chem. Soc. A 1966, 580.
195) Leach, J. B., Morris, J. H.: Organometal. Chem. Rev. A 4, 1 (1969).
196) — — J. Organometal. Chem. (Amsterdam) 13, 313 (1968).
197) Nöth, H., Regnet, W.: Advan. Chem. Ser. 42, 166 (1964).
198) Lappert, M. F., Majumdar, M. K., Tilley, B. P.: J. Chem. Soc. A 1966, 1590.
199) Mikhailov, B. M., Kozminskaya, T. K.: Izv. Akad. Nauk. SSSR, Ser. Khim. 1965, 439.
200) Niedenzu, K., Beyer, H., Dawson, J. W.: Inorg. Chem. 1, 738 (1962).
201) Nöth, H., Regnet, W.: Chem. Ber. 102, 167 (1969).
202) Miller, J. J., Johnson, F. A.: J. Am. Chem. Soc. 90, 218 (1968).
203) Thomas, P. C., Paul, I. C.: Chem. Commun. 1968, 1130.
204) Nöth, H., Regnet, W.: Chem. Ber. 102, 2241 (1969).
205) Fritz, P. W., Niedenzu, K., Dawson, J. W.: Inorg. Chem. 4, 886 (1964).

Received November 10, 1969

1,3,2-Diazaboracycloalkanes

Prof. Dr. K. Niedenzu and C. D. Miller, M. A.

Department of Chemistry, University of Kentucky, Lexington, KY. 40506, USA

Contents

A. Introduction 191

B. Syntheses 192

C. The Structure of 1,3,2-Diazaboracycloalkanes 197

D. Chemical Reactions 199

E. Physicochemical Studies 202

F. Conclusion 204

G. References 204

A. Introduction

Sigma-bonded boron-nitrogen-carbon heterocycles were first described in 1955 [1] and since that time a substantial volume of research has been carried out on such systems. Two major streams of investigations have evolved.

Most attention has been devoted to an exploration of the chemistry of the *borazaromatics*. In these compounds, two neighboring carbon atoms of an aromatic system are replaced by the isoelectronic boron-nitrogen groups as illustrated by the 2,1-borazaronaphthalene, *1* [2,3]. Their great chemical stability is a distinct feature of heteroaromatic materials and several reviews of their chemistry have been published [4-6].

The second major type of sigma-bonded boron-nitrogen-carbon heterocycles contains boron and nitrogen *as annular atoms of saturated ring systems*. Though several types of such *azaboracycloalkanes* are known [7] the present article will be concerned only with the 1,3,2-diazaboracylo-alkane system, 2.

Only in more recent years have these compounds received some detailed attention. Their chemistry was studied primarily in order to elucidate the nature of bonding between boron and nitrogen in such sigma-bonded heterocyclic systems. However, investigations are in progress in order to evaluate the potential of 1,3,2-diazaboracycloalkanes in cancer therapy and the utilization of boron containing heterocycles as insecticides, fungicides and antiseptics has been patented for several years [8,9]. A general feature of the 1,3,2-diazaboracycloalkanes appears to be their *great thermal stability*. Also, recent studies [10] suggests that exocyclic substitution may greatly enhance the hydrolytic stability of the 1,3,2-diazaboracycloalkane system and this event may tend to stimulate further interest in these materials.

B. Syntheses

So far three primary routes for the synthesis of 1,3,2-diazaboracyclo-alkanes have been developed.

The original procedure as reported by Goubeau and Zappel in their pioneering work [1] involves the pyrolysis of a mixture of trialkylborane and aliphatic α,ω-diamines. It was shown that the reaction proceeds through various intermediates and it can be illustrated by the following Eqs. (1)—(3).

$$B(CH_3)_3 + H_2N-(CH_2)_2-NH_2 \xrightarrow[\text{temperature}]{\text{room}} \underset{3}{H_2N-(CH_2)_2-NH_2 \cdot B(CH_3)_3} \quad (1)$$

$$\underset{3}{} \xrightarrow{250\,°C} CH_4 + \underset{4}{H_2N-(CH_2)_2-NH-B(CH_3)_2} \quad (2)$$

$$\underset{4}{} \xrightarrow{475\,°C} CH_4 + \begin{array}{c} H_2C\text{———}CH_2 \\ | \qquad\quad | \\ HN \qquad NH \\ \diagdown \;\; B \;\; \diagup \\ | \\ CH_3 \end{array} \quad (3)$$

It is quite possible that the intermediate 4 has a coordinated ring structure 5 and thereby is favoring the elimination of additional methane to yield the all-sigma-bonded ring system 2.

$$
\begin{array}{c}
\text{H}_2\text{C}\!-\!\!-\!\!\text{CH}_2 \\
| \qquad | \\
\text{HN}\diagdown_{\text{B}}\diagup\text{NH}_2 \\
\text{H}_3\text{C}\diagup \quad \diagdown\text{CH}_3
\end{array}
$$

5

Pailer and Fenzl [11] replaced the trialkylboranes in this pyrolysis reaction by organoboron dihalides, RBX_2, and obtained the 1,3,2-diazaboracycloalkane system under evolution of hydrogen halide.

However, the known difficulties in performing controlled pyrolysis reactions in a laboratory prompted a search for more manageable synthetic procedures. These studies resulted in a detailed investigation of transamination reactions [10,12-15] which have since become one of the most useful methods for the preparation of B-organosubstituted 1,3,2-diazaboracycloalkanes (Eq. 4).

$$
RB\!\!\begin{array}{c}\diagup\text{N(CH}_3)_2 \\ \diagdown\text{N(CH}_3)_2\end{array} + \begin{array}{c}\text{HNR}'\!\!-\!\!\!- \\ \quad\quad (\text{CH}_2)_n \\ \text{HNR}'\!\!-\!\!\!\!-\end{array} \rightarrow 2\,\text{HN(CH}_3)_2 + RB\!\!\begin{array}{c}\diagup\text{NR}'\!\!-\!\!\!- \\ \quad\quad (\text{CH}_2)_n \\ \diagdown\text{NR}'\!\!-\!\!\!\!-\end{array} \quad (4)
$$

A stepwise mechanism has been suggested for this reaction [21] whereby a coordinated ring system analogous to *5* is formed first and which subsequently looses a second molecule of dimethylamine to form the final product.

A third basic procedure of synthesizing 1,3,2-diazaboracyloalkanes involves the interaction of aliphatic α,ω-diamines with boron trihalides in the presence of a tertiary amine as hydrogen halide acceptor as is illustrated by Eq. 5 [15,17].

$$
BX_3 + 2\,NR_3 + \begin{array}{c}\text{HNR}'\!\!-\!\!\!- \\ \quad\quad (\text{CH}_2)_n \\ \text{HNR}'\!\!-\!\!\!\!-\end{array} \rightarrow 2\,\text{HNR}_3X + XB\!\!\begin{array}{c}\diagup\text{NR}'\!\!-\!\!\!- \\ \quad\quad (\text{CH}_2)_n \\ \diagdown\text{NR}'\!\!-\!\!\!\!-\end{array} \quad (5)
$$

More recently, it was found that the experimental procedure can be improved by originating from the trialkylamine-trihaloboranes, $R_3N \cdot BX_3$, rather than from the boron trihalide [18].

In addition to pyrolysis, transamination, and dehydrohalogenation with tertiary amines several other reactions have been described which yield access to the 1,3,2-diazaboracycloalkane system. For example, Abel and Bush [19] reported the ready replacement of silicon by boron in certain heterocyclic systems as illustrated in Eq. (6). It seems quite likely that this procedure can be developed into a general synthesis. On the other hand, the interaction of trialkylboranes with bis(1,3-diphenylimidazolidinylidene-2) as described by Hesse and Haag [20] according to Eq. (7) appears to be of limited value.

$$
\begin{array}{c}
H_2C-NR \\
| \\
H_2C-NR
\end{array}
Si
\begin{array}{c}
CH_3 \\
\\
CH_3
\end{array}
+ C_6H_5BCl_2 \rightarrow (CH_3)_2SiCl_2 + C_6H_5B
\begin{array}{c}
NR-CH_2 \\
| \\
NR-CH_2
\end{array}
\qquad (6)
$$

$$
\begin{array}{c}
H_2C-NR \\
| \\
H_2C-NR
\end{array}
C=C
\begin{array}{c}
NR-CH_2 \\
| \\
NR-CH_2
\end{array}
+ 2\,BR'_3 \rightarrow 2\,R'_3C-B
\begin{array}{c}
NR-CH_2 \\
| \\
NR-CH_2
\end{array}
\qquad (7)
$$

The Tables 1—3 list 1,3,2-diazaboracycloalkanes of type *2* in dependence of their ring size. The five- and six-membered ring systems are the most common though several seven-membered rings have been described. Apparently no detailed effort has been made so far to synthesize larger ring systems. It has been suggested [21] that, for steric reasons, the formation of eight- or nine-membered rings may be suppressed by the formation of linear molecules though it may be possible to synthesize ring systems with more than eleven annular atoms.

Table 1. *1,3,2-Diazaboracyclopentanes 2 (n = 2)*

R	R'	R''	m.p.,°C	b.p. (mm),°C	Ref.
CH_3	H	H	43.5	106	1,12)
CH_3	CH_3	CH_3		124 (738)	18)
CH_3	C_2H_5	C_2H_5		33—34 (2)	10)
$CHCH_2$	CH_3	CH_3		35—36 (11)	14)
OCH_3	CH_3	CH_3		49—51 (12)	17)
OCD_3	CH_3	CH_3		44 (10)	17)
Cl	CH_3	CH_3		43—45 (12)	17)
Cl	C_2H_5	C_2H_5		65—66 (10)	16)
Cl	$i-C_3H_7$	$i-C_3H_7$		64 (3.5)	16)
$n-C_4H_9$	H	CH_3		33—35 (2)	10)
$n-C_4H_9$	CH_3	CH_3		31—32 (2)	10)
$n-C_4H_9$	H	C_2H_5		40—43 (2)	10)
$n-C_4H_9$	C_2H_5	C_2H_5		45—47 (2)	10)
$t-C_4H_9$	C_6H_5	C_6H_5	35—37		20)
$C(n-C_4H_9)_3$	C_6H_5	C_6H_5	64—65		20)
$C(CH_2-C_6H_5)_3$	C_6H_5	C_6H_5	97—99		20)
C_6H_5	H	H	157		11)
C_6H_5	H	CH_3		66—68 (2)	10)
C_6H_5	H	C_2H_5		70—72 (2)	10)
C_6H_5	CH_3	CH_3		73 (3)	13)
C_6H_5	C_2H_5	C_2H_5		95 (8)	13,19)
$N(CH_3)_2$	CH_3	CH_3		33—34 (1)	18)

Table 2. *1,3,2-Diazaboracyclohexanes 2 (n = 3)*

R	R'	R''	m.p.,°C	b.p. (mm),°C	Ref.
H	H	H		97—99	[15]
CH_3	H	H		36—38 (20)	[1,15]
CH_3	H	C_2H_5		39—41 (1.5)	[10]
CH_3	CH_3	CH_3		30—31 (2)	[10]
$CHCH_2$	H	H		41 (7)	[15]
C_2H_5	H	H		44—45 (13)	[15]
$n-C_4H_9$	H	CH_3		44—46 (2)	[10]
$n-C_4H_9$	H	C_2H_5		48—50 (2)	[10]
$n-C_4H_9$	CH_3	CH_3		48—51 (2)	[10]
C_6H_5	H	H	50—52	94—94 (1)	[15]
C_6H_5	H	CH_3		74—75 (2)	[15]
C_6H_5	CH_3	CH_3		71—74 (2)	[15]
C_6H_5	H	C_2H_5		81—83 (2)	[15]
Cl	CH_3	CH_3		64 (10)	[16]
$N(CH_3)_2$	CH_3	CH_3		50 (1)	[18]
Br	CH_3	CH_3		53—57 (7)	[18]
F	CH_3	CH_3		72° (45)	[18]
C_6H_5	H	$(CH_2)_3-NH_2$		122° (1)	[21]
C_2H_5	H	$(CH_2)_3-NH-B(C_2H_5)_2$		97—97 (1)	[21]

Table 3. *1,3,2-Diazaboracycloheptanes 2 (n = 4)*

R	R'	R''	m.p.,°C	b.p. (mm),°C	Ref.
CH_3	H	H		39—42 (6)	[15]
CH_3	CH_3	CH_3		34—35 (2)	[10]
CH_3	C_2H_5	C_2H_5		43—45 (2)	[10]
$n-C_4H_9$	H	H		64 (1)	[15]
$n-C_4H_9$	CH_3	CH_3		52—54 (2)	[10]
$n-C_4H_9$	C_2H_5	C_2H_5		65—68 (2)	[10]
C_6H_5	H	H	27—29	102—104 (1)	[15]
C_6H_5	CH_3	CH_3		85—86 (2)	[10]
C_6H_5	C_2H_5	C_2H_5		112—114 (3)	[10]

Besides the 1,3,2-diazaboracycloalkanes of type *2* as listed in Tables 1—3 it is possible to synthesize heterocycles with substituents on the methylene bridges. For example, transamination reactions have been performed with N,N',2-trimethyl-diaminopropane and the following compounds of type *6* were isolated in good yield [18].

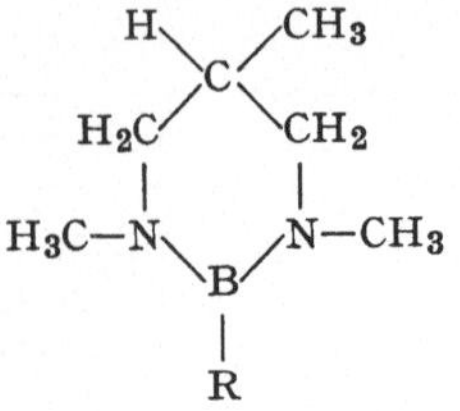

R	b.p. °C (mm)
$N(CH_3)_2$	81 (14)
F	49 (16)

6

Also, several polycyclic 1,3,2-diazaboracycloalkane derivatives have been reported. The first of these was the 1,8,10,9-triazaboradecalin, 7 [15,21], which is readily obtained from a transamination reaction between tris(dimethylamino)borane and 3,3′-diaminodipropylamine as depicted in Eq. (8).

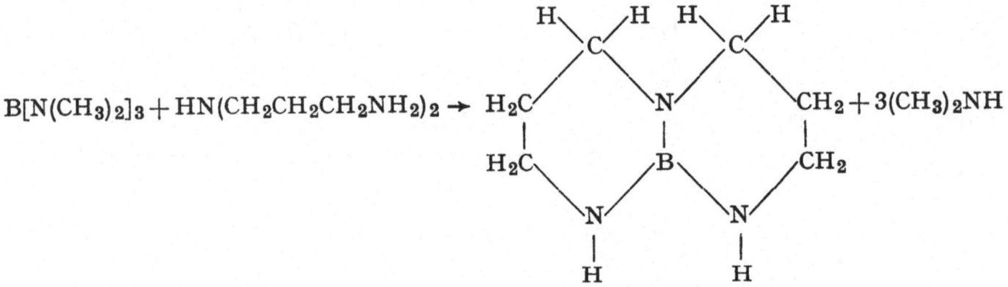

$$B[N(CH_3)_2]_3 + HN(CH_2CH_2CH_2NH_2)_2 \rightarrow$$

$$+ 3(CH_3)_2NH \quad (8)$$

7

m.p. 39—41 °C
b.p. 62 °C (1 mm)

Several derivatives of 7 have been described [22]. Also, 1,7,8,9-triazaborahydrindan, *8*, has been obtained in similar fashion from tris(dimethylamino)borane and N-aminoethyl-1,3-diaminopropane [23].

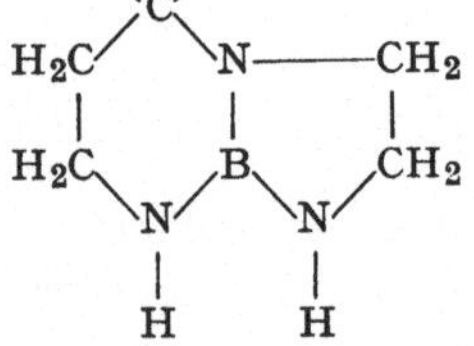

8

b.p. 79 °C (2 mm)

C. The Structure of 1,3,2-Diazaboracycloalkanes

The nature of the boron-nitrogen sigma bond has been subject of many investigations [36]. Heterocyclic molecules are particularly suitable systems for studying the bonding between boron and nitrogen. This feature has been a primary driving force in the studies of 1,3,2-diazaboracycloalkanes.

On the basis of the Raman spectrum of 2-methyl-1,3,2-diazaboracyclohexane Goubeau and Zappel [1] concluded that the boron-nitrogen bonding in such heterocycles is similar to that of borazine, $(-BH-NH-)_3$. This reasoning was based primarily on the observation of the NH stretching modes near 3465 cm^{-1} and of the antisymmetric BN$_2$ valence vibration near 1490 cm^{-1}. A detailed discussion of the vibrational spectrum of the cited compound [25] including a study of isotopically labeled derivatives of the material confirmed the basic findings. Additional spectroscopic studies [21,24] substantiated the concept of a relatively high BN bond order in the heterocyclic 1,3,2-diazaboracycloalkane systems. This assumption is based primarily on the observation of the antisymmetrical BN valence vibration near 1500 cm^{-1} which is then related directly to the bond order. However, force constants have not yet been calculated and it may be possible that frequency coupling may lead to abnormally high values for the BN stretching mode thereby indicating a higher BN bond order than actually exists.

If one assumes participation of the pi-electrons of nitrogen in the boron-nitrogen bonding according to

$$\overset{..}{>}N-B< \ \longleftrightarrow \ >N\text{-}\text{-}\text{-}B<$$

such an event is possible only in the case of planar or nearly planar bonding arrangement of nitrogen. Due to sp^2 hybridization all nitrogen substituents should be coplanar. On that basis, the molecular fragment 9 of 1,3,2-diazaboracycloalkanes should be coplanar and thereby restrict the geometry of the molecules. A heterocycle of type 2 with $n = 2$ should

9

be completely planar whereas in the case of *2, n = 3*, only the center carbon atom should be distorted from the molecular plain.

A detailed study of the proton magnetic resonance spectra of several 1,3,2-diazaboracycloalkanes [26)] has given evidence for the planarity of the ring system. In the temperature range from −70 °C to +90 °C no changes in the appearance of the basic spectra were observed; an example is illustrated in the following figure, depicting the proton magnetic resonance spectrum of 1,3-dideuterio-2-phenyl-diazaboracyclohexane.

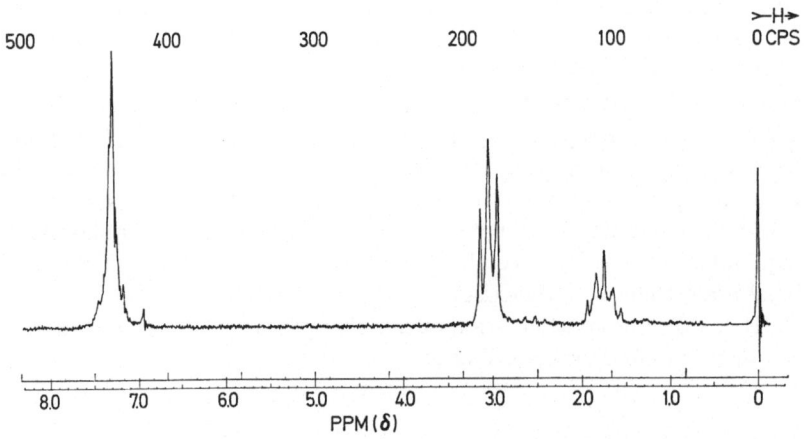

[1]H n.m.r. Spectrum of 1,3-Dideutero-2-phenyl-diazaboracyclohexane (solution in CDCl$_3$ at 38 °C, external TMS reference)

Also, no solvent or concentration effect has been noted and a spectrum calculated for a planar ring system was consistent with the experimental data. The coupling constant J_{HNCH} in boron-substituted 1,3,2-diazaboracyclohexanes was found to be about 3 c.p.s. indicating a planar valence arrangement of the nitrogen. These cited results tend to support the assumption of a BN bond order greater than 1.

A crystal structure determination [27)] of 1,8,10,9-triazaboradecalin, *7*, has shown that in this compound the part *10* of the molecule is nearly flat with a greatest deviation from planarity being 0.052 Å for the nitrogen atom 10.

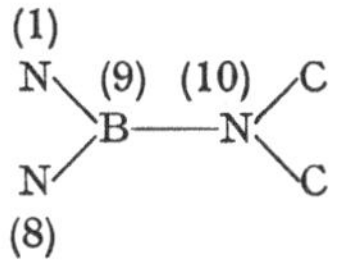

(1)
N   (9) (10)   C
 \ \B——N/   /
  \ /   \ \
  N/   \C
(8)

10

The molecular shape of the entire molecule is closest to that of a distorted naphthalene. The configuration at the boron atom 9 is almost exactly trigonal planar but that at the nitrogen atom 10 is very slightly pyramidal. The essentially planar configurations of the nitrogen atoms, the coplanarity of the six central atoms, and the short B—N bond lengths with 1.41 to 1.43 Å indicate pi-bonding superimposed on the sigma-bonding. Hence these results tend to substantiate the vibrational data of a high BN bond order in such heterocycles. This finding is supported by some simplified molecular orbital calculations on the 1,8,10,9-triazaboradecalin which resulted in a BN bond order of about 1.31 [27].

D. Chemical Reactions

Very little is yet known about the specific chemistry of 1,3,2-diazaboracycloalkanes. At room temperature, these compounds add hydrogen halide readily in a 1:2 molar ratio [1]. In a similar manner, boron trihalide is added at low temperatures [13]. The solid addition products of the latter reaction decompose near room temperature with cleavage of the original B—N linkages presumably leading to linear products (Eq. 9).

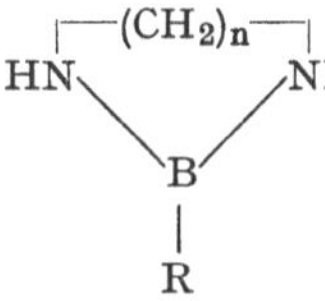

$$\text{HN} \underset{\underset{R}{\overset{|}{B}}}{\overset{\overset{(CH_2)_n}{\frown}}{\diagdown\diagup}} \text{NH} + 2\, BX_3 \rightarrow X_2B-NH-(CH_2)_n-NH-BX_2 + RBX_2 \qquad (9)$$

Unless the nitrogen is protected by an organic exocyclic substituents, 1,3,2-diazaboracycloalkanes with reactive groups at the boron atom tend to undergo intermolecular condensation reactions. For example, 2-dimethylamino-1,3,2-diazaboracycloalkanes have not yet been isolated in pure form. Rather, condensation to the borazine derivative *11* seems to occur at relatively low temperatures [21] as illustrated in Eq. 10.

$$3 \quad \begin{array}{c} H \\ H_2C \end{array} \begin{array}{c} C \\ \end{array} \begin{array}{c} H \\ CH_2 \end{array} \longrightarrow \qquad \qquad (10)$$

11

m.p. 154—155 °C
b.p. 190—195 °C (1 mm)

However, 1,3-diorgano-2-dimethylamino-diazaboracycloalkanes have been prepared by a simple transamination reaction between tris(dimethylamino)borane, $B[N(CH_3)_2]_3$, and α,ω-diorgano-polymethylenediamines [29].

Chlorine bonded to the boron atom of 1,3,2-diazaboracycloalkanes has been replaced by other halogen atoms in simple transhalogenation reactions [29]. Also, dimethylamino groups can be replaced by halogen in a ligand exchange reaction, for example, with boron trifuloride-etherate. However, it should be stressed, that in all these reactions the nitrogen atoms of the heterocycle must be protected by exocyclic organic substituents.

2-halo-1,3,2-diazaboracycloalkanes readily interact with Grignard reagents (Eq. 11).

$$R'MgX + XB \underset{NR}{\overset{NR}{\diagup}} (CH_2)_n \rightarrow MgX_2 + R'B \underset{NR}{\overset{NR}{\diagup}} (CH_2)_n \qquad (11)$$

Preliminary experiments [10] have shown that 1,3,2-diazaboracycloalkanes can be dehydrogenated to yield derivatives containing carbon-carbon double bonds. The first characterized products of this type, however, 1,3-dimethyl-2-n-butyl-diazaboracycloheptene(5), b.p. 59—61°C (2 mm), and 1,3,2-trimethyldiazaboracycloheptene(5), b.p. 43—44 °C (2 mm), were obtained by a transamination reaction [10].

The chemistry of 1,8,10,9-triazaboradecalin, 7, has been explored in considerable detail. The 1,8-dilithio compound is readily accessible [22] and offers interesting preparative aspects. Originating from the dilithio compound, the novel heterocyclic systems *12* [30] and *13* [22] have been prepared. Also, the dilithio compound was utilized for the synthesis of unsymmetrically substituted borazines [22].

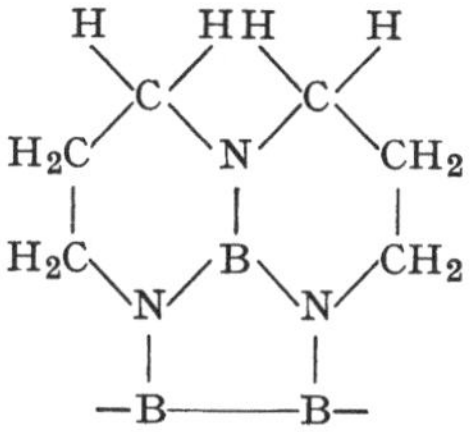

12

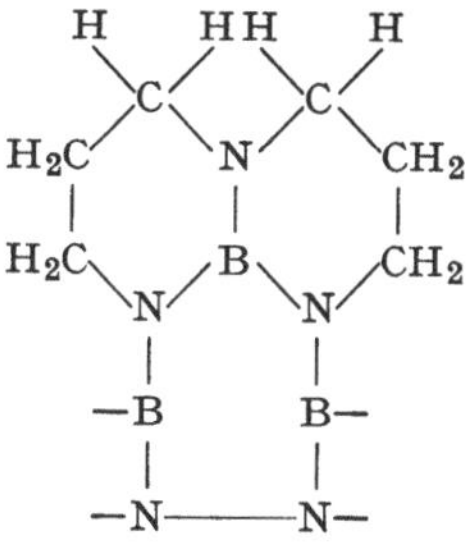

13

Insertion of isocyanate in the 1,8,10,9-triazaboradecalin [22] has been found to yield *14* (Eq. 12). In contrast, phenyl isothiocyanate was found

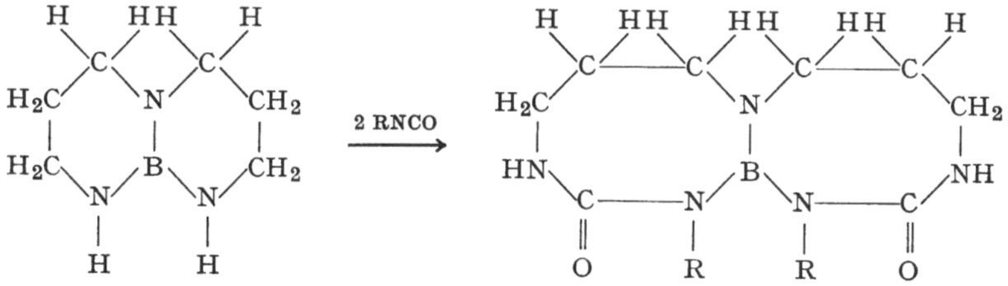

14   m.p. 175 °C

to interact with 1,8,10,9-triazaboradecalin in a 1:1 molar ratio only. The resultant product has been formulated as shown in structure *15*.

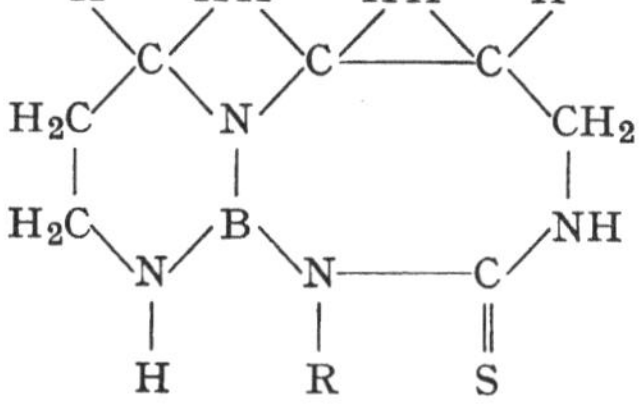

15

m.p. 110—112 °C

E. Physicochemical Studies

As outlined above, vibrational spectroscopic studies have been used largely in order to elucidate the nature of the bonding between boron and nitrogen in 1,3,2-diazaboracycloalkanes by relating BN stretching modes with the bond order. It is worth noting, however, that the frequency of the antisymmetrical BN valence vibration appears to be relatively independent of the electronic nature of exocyclic substituents at either the nitrogen or the boron atom [28].

A screening of the vibrational spectra of 1,3,2-diazaboracycloalkanes has indicated a relationship between the ring size and the frequency of a ring pulsation mode [28]. It was noted that the frequency position of the ring pulsation appears to reflect the ring size (Table 4) but does not give an indication about ring strain. This observation correlates well with the situation noted for cycloalkanes.

Table 4. *Ring pulsation frequencies* (cm^{-1})

Number of annular atoms	Monomethylcyccloalkanes	2-Methyl-1,3,2-diaza-boracycloalkanes
5	890	858
6	770	763
7	720	713

Very few boron-11 nuclear magnetic resonance studies have been reported on 1,3,2-diazaboracycloalkanes though p.m.r. spectra have been utilized in elucidating the structure of several of these heterocycles [15,19,26,34]. 1,3,2-diazaboracyclohexane, *16*, is the sole representative of these boron-nitrogen-carbon ring systems in which the exocyclic boron substituent constitutes a hydrogen atom [15]. It is noteworthy that the coupling constant J_{BH} of this compound compares favorably with J_{BH} of borazine, $(-BH-NH-)_3$, and of N-trimethylborazine, $(-BH-NCH_3-)_3$ [15,31].

16

So far, only one detailed discussion of boron-11 nuclear magnetic resonance spectra of aminoborane systems has been reported [31]. It was found that the [11]B chemical shifts of aminoborane systems can be described fairly well in terms of a set of additive substituent contributions. In consonance with earlier work on trisubstituted boron compounds [35] these contributions depend on the mesomeric effects of substituents rather than their electronegativity. 1,3,2-diazaboracycloalkanes can be considered as aminoborane derivatives and in the case of the known heterocycles the exocyclic boron substituent will govern primarily the boron chemical shifts and will do so by mesomeric effects. However, the available data are rather limited and it may be possible that additional factors must be considered. Steric effects appear to be negligible, however, since the heterocycles with either six or seven annular atoms have almost identical shifts (Table 5).

Table 5. ^{11}B *chemical shifts of 2-substituted 1,3,2-diazaboracycloalkanes, $RB(NH)_2(CH_2)n$*

n	R	Chemical shift (p.p.m., from trimethylborate)
3	H	-7.1
3	CH_3	-11.8
3	C_2H_5	-12.0
3	C_6H_5	-9.0
3	$CHCH_2$	-8.9
4	CH_3	-11.8
4	$n-OC_4H_9$	-12.5
4	C_6H_5	-10.4

Mass spectroscopic studies on 1,3,2-diazaboracycloalkanes [18] seem to reveal some interesting aspects. It appears that dehydrogenation at annular carbon atoms under electron impact is readily accomplished. This observation tends to indicate that the resultant products may be stable and suggests research in this area. Moreover, it is noteworthy that in all the spectra which have been studied so far, no complete breakdown of the molecules was observed. Instead, species with one of the two original boron-nitrogen bonds left intact appear to be rather stable and there seems to be a particular tendency for the intramolecular formation of boron-containing species with one B—N and at least one B—C bond. Also, a characteristic breakdown pattern in all of the studied spectra of 1,3,2-diazaboracycloalkanes seems to be established inasmuch as the ready loss of exocyclic groups other than those attached to the boron atom has been observed.

Dewar and Rona [37] have reported the appearance of very intense peaks due to doubly charged ions in the mass spectrum of 1,3-dimethyl-2-phenyl-diazaboracyclopentane. The ready formation of dipositive ions at relatively low ionizing potentials was interpreted to indicate an unusual stability of the ions.

F. Conclusion

Within less than two decades after the discovery of sigma-bonded boron-nitrogen-carbon heterocycles a sizeable volume of research has been carried out on such compounds. In the area of 1,3,2-diazaboracyclo-alkanes a variety of preparative procedures has been explored and, on the basis of spectroscopic studies, some insight into the bonding situation in such molecules has been gained. It is reasonably well established that the bonding between boron and nitrogen involves a sizeable degree of double bond character.

1,3,2-diazaboracycloalkanes are thermally quite stable, but in general, hydrolize quite readily. However, it appears that the hydrolytic stability can be improved by exocyclic substitution though it has not yet been established what factors are responsible for such an improvement.

Evidence has been presented that 1,3,2-diazaboracycloalkanes can be dehydrogenated and hence they may become valuable precursors for the synthesis of heteroaromatic systems. In this connection, the 1,3,2-diazaboracyclopentane appears to be of particular scientific interest. Dehydrogenation can yield a system which is isoelectronic with the cyclopentadienyl anion. Such a system could open an area of additional fruitful research.

So far, practical applications of 1,3,2-diazaboracycloalkanes appear to be rather limited. However, the recent utilization of an aromatic boron derivative as a tranquilizer [32,33] may have an important bearing on increasing the pharmocodynamic studies of boron compounds for clinical uses in general. It remains to be seen whether or not this event will stimulate a wider interest in 1,2,3-diazaboracycloalkanes.

G. References

[1] Goubeau, J., Zappel, A.: Z. Anorg. Allgem. Chem. 279, 38 (1955).
[2] Dewar, M. J. S., Dietz, R.: J. Chem. Soc. 1959, 2728.
[3] — — Kubba, V., Lepley, A. R.: J. Am. Chem. Soc. 83, 1754 (1961).
[4] — In: Progress in Boron Chemistry, Vol. I, p. 235, 1964.
[5] — Advan. Chem. Ser. 42, 227 (1964).
[6] Maitlis, P. M.: Chem. Rev. 62, 223 (1962).

[7] Niedenzu, K.: Allgem. Prakt. Chem. *17*, 596 (1966).
[8] Conklin, G. W., Morris, R. C.: Brit. Pat. 760,090 (1958).
[9] Garner, G. J.: U. S. Pat. 2,839,654 (1958).
[10] Weber, W., Dawson, J. W., Niedenzu, K.: Inorg. Chem. *5*, 726 (1966).
[11] Pailer, M., Fenzl, W.: Monatsh. Chem. *92*, 1294 (1961).
[12] Nöth, H.: Z. Naturforsch. *16b* 470 (1961).
[13] Niedenzu, K., Beyer, H., Dawson, J. W.: Inorg. Chem. *1*, 738 (1962).
[14] Fritz, P., Niedenzu, K., Dawson, J. W.: Inorg. Chem. *3*, 626 (1964).
[15] Niedenzu, K., Fritz, P., Dawson, J. W.: Inorg. Chem. *3*, 1077 (1964).
[16] Brown, M. P., Dann, A. E., Hunt, D. W., Silver, H. B.: J. Chem. Soc. *1962*, 4648.
[17] Meller, A., Maracek, H.: Monatsh. Chem. *98*, 2336 (1967).
[18] Niedenzu, K., and coworkers: unpublished data.
[19] Abel, E. W., Bush, R. P.: J. Organometal. Chem. *3*, 245 (1965).
[20] Hesse, G., Haag, A.: Tetrahedron Letters *16*, 1123 (1965).
[21] Niedenzu, K., Fritz, P.: Z. Anorg. Allgem. Chem. *340*, 329 (1965).
[22] Fritz, P., Niedenzu, K., Dawson, J. W.: Chem. *4*, 886 (1965).
[23] Niedenzu, K., Weber, W.: Z. Naturforsch. *21b*, 811 (1966).
[24] Goubeau, J., Schneider, H.: Liebigs Ann. Chem. *675*, 1 (1964).
[25] Dawson, J. W., Fritz, P., Niedenzu, K.: J. Organometal. Chem. *5*, 211 (1966).
[26] Niedenzu, K., Miller, C. D., Smith, S. L.: Z. Anorg. Allgem. Chem. in press.
[27] Bullen, G. J., Clark, N. H.: J. Chem. Soc. *1969A*, 404.
[28] Niedenzu, K., Fritz, P.: J. Anorg. Allgem. Chem. *344*, 329 (1966).
[29] — Busse, P. J.: unpublished.
[30] Nöth, H., Abeler, G.: Chem. Ber. *101*, 969 (1968).
[31] Scott, K. N., Brey, W. S.: Inorg. Chem. *8*, 1703 (1969).
[32] De Boislambert, P., De Wanger, G.: Arch. Hosp. (Paris) *32*, 106 (1960).
[33] Trabucchi, C., Zuanazzi, J. F.: Gazz. Med. Ital. *119*, 223 (1960).
[34] Niedenzu, K., Dawson, J. W., Fritz, P.: Z. Anorg. Allgem. Chem. *342*, 297 (1966).
[35] Nöth, H., Vahrenkamp, H.: Chem. Ber. *99*, 1049 (1966).
[36] Niedenzu, K., Dawson, J. W.: Boron-Nitrogen Chemistry. Berlin-Heidelberg-New York: Springer 1965.
[37] Dewar, M. J. S., Rona, P.: J. Am. Chem. Soc. *87*, 5510 (1965).

Received November 10, 1969

Darstellung und Systematisierung von Boraten und Polyboraten

Prof. Dr. G. Heller

Institut für Anorganische Chemie der Freien Universität Berlin

Inhalt

1.   Einleitung 206
1.1. Bildung und Vorkommen von Polyboraten 208
1.2. Zur Systematik 208
1.3. Bindungsprinzipien in Boraten 209
1.4. Möglichkeiten zur Strukturaufklärung............................ 210
1.5. Strukturen von Polyboraten in Lösung 211

2.   Experimentelles 213

3.   Ergebnisse und Diskussion 215
3.1. Ammoniumpolyborate 215
3.2. Lithiumborate 218
3.3. Natriumborate 219
3.4. Kaliumborate 225
3.5. Rubidium- und Caesiumborate 237
3.6. Beryllium- und Magnesiumborate .....;........................... 240
3.7. Calcium-, Strontium- und Bariumborate........................... 243
3.8. Andere Borate mit Metallkationen 248
3.9. Polyborate mit organischen Kationen 249

4.   Schema zur Systematik von Boratstrukturen 262
4.1. Monoborate und Monoborat-Polymere 264
4.2. Diborate und Diborat-Polymere 266
4.3. Triborate und Triborat-Polymere 266
4.4. Tetraborate und Tetraborat-Polymere 269
4.5. Pentaborate 271
4.6. Höhere Borate 273

5.   Literatur... 276

1. Einleitung

Es wird eine Zusammenstellung neuerer und auch eigener Untersuchungen gegeben, um derart einen Überblick über die strukturellen Zusammenhänge zwischen den verschiedenen Polyborat-Typen zu erhalten. Für die eigenen Untersuchungen wurden zwei moderne Methoden benutzt, die es erlauben, Aussagen über Strukturen von Polyboraten zu machen, ohne die in der Boratchemie besonders schwierigen Röntgenstrukturanalysen durchführen zu müssen:

1. Darstellung von Polyboraten durch Hydrolyse von Borsäureestern in *organischen Lösungsmitteln* in Gegenwart von Basen,

2. die quantitative Auswertung von *Protonenresonanzmessungen*.

Durch *Hydrolyse von Borsäureestern* in Gegenwart von Basen wurden je nach den Konzentrationsverhältnissen verschiedene Polyborate mit den Kationen Kalium, Caesium, Ammonium, Guanidinium, Piperidinium, Tetraalkyl- und Trialkylammonium erhalten. Ein großer Teil dieser Verbindungen war bisher unbekannt und wurde durch chemische Analysen, IR- und Remissionsspektren sowie durch Röntgendaten charakterisiert. Einige der Verbindungen waren schon früher aus wäßriger Lösung dargestellt worden. Die Base: Säure-Verhältnisse der aus organischem Lösungsmittel gewonnenen Salze entsprechen im großen und ganzen denen der aus wäßriger Lösung erhaltenen Polyborate. Bei den mit organischen stickstoffhaltigen Kationen dargestellten Verbindungen handelt es sich ebenso wie bei Ammonium vorwiegend um 1:5-Borate; auch das Kalium (1:5)-Borat ist sowohl in Wasser als auch in organischen Lösungsmitteln das am schwersten lösliche Salz. Abweichend davon bildet Guanidinium bei Darstellung nach der Esterhydrolysenmethode basereichere Polyborate, während mit den sperrigen Tetra- und Trialkylammonium-Kationen baseärmere 1:7-Borate entstehen, die bisher ebenfalls unbekannt waren. Molekulargewichtsbestimmungen mit Hilfe der Ultrazentrifuge zeigen, daß sie in verdünnter Lösung im organischen Medium wie in Wasser in monomere Teilchen zerfallen. Die Existenz der „Hexaborate" konnte nicht bestätigt werden. Eines der dargestellten Kaliumborate besitzt das ungewöhnliche Base: Säure-Verhältnis 1:12; mit Hilfe von PMR-Messungen konnte gezeigt werden, daß es sich tatsächlich um ein Salz handelt, und nicht etwa um Kaliumionen, die an Borsäure absorbiert sind. Mit konz. wäßrigen Ammoniak wurde ein völlig wasserfreies Ammoniumpolyborat $\{NH_4B_3O_5\}_n$ erhalten, das gummiartig anfällt und hochkondensiert sein muß.

Die Methode der Boresterhydrolyse ist nicht unbeschränkt anwendbar: in Gegenwart schwächerer Basen wie Pyridin oder Urotropin wurde nur Borsäure erhalten.

Die Darstellung von Polyboraten durch Esterhydrolyse in einem organischen Lösungsmittel hat über die präparative Bedeutung hinaus einen großen Vorteil: erfahrungsgemäß sind nämlich alle so gewonnenen Verbindungen kristallwasserfrei. Am Beispiel zweier Kaliumpolyborate wurde dies mit Hilfe von *Protonenresonanzmessungen* bewiesen. Da es bei den aus wäßriger Lösung dargestellten Polyboraten schwierig ist, zwischen Kristall- und Strukturwasseranteil zu unterscheiden, werden in der Literatur oft nur die Summenformeln m $M_2O \cdot n$ $B_2O_3 \cdot x$ H_2O angegeben. Weiß man aber, daß die Verbindung nur Strukturwasser enthält, so kann man gemäß $M_{2m}[B_{2n}$ $O_{m+3n-x}(OH)_{2x}]$ die Strukturformel aufstellen und den Strukturtyp diskutieren.

Aufgrund von Protonenresonanzmessungen an Natriumtetraboraten und an Kaliummono-, -tetra- und -pentaboraten, die aus wäßrigen Lösungen erhalten wurden, sowie an Produkten, die aus diesen Verbindungen durch thermischen Abbau entstehen, wurden ebenfalls deren Strukturformeln aufgestellt. Dabei wurden in der Literatur aufgetretene Widersprüche über Polyboratstrukturen geklärt. Wie aus den ersten Ableitungen der PMR-Kurven bei 90 °K deutlich wird, ist für das Mineral Tinkalkonit die Formel $Na_2(H_2O)_3[B_4O_5(OH)_4]$ und für das Kaliummonoborat $K_2O \cdot B_2O_3 \cdot 2,67$ H_2O die Formel eines Schichtenborats, $\{[K(H_2O)]_3$-$[B_3O_5(OH)_2]\}_n$, zu schreiben. Es wurde gezeigt, daß $K[B_5O_6(OH)_4]$ nicht durch thermischen Abbau von $K(H_2O)_2[B_5O_6(OH)_4]$ zu erhalten ist, wohl aber durch Boresterhydrolyse.

Von den bekannten, durch Kristallstrukturanalyse gesicherten Boratstrukturen ausgehend, wurde unter Berücksichtigung der neuen experimentellen Untersuchungen für eine große Anzahl von Polyboratstrukturen eine Systematik aufgestellt. Dabei wurde versucht, alle denkbaren Mono-, Di-, Tri-, Tetra- und Pentaborate sowie ihre Polymere zu berücksichtigen.

1.1. Bildung und Vorkommen von Polyboraten

Polyborate sind die Salze der Isopolysäuren des Bors. Isopolysaure Salze bilden sich durch Wasseraustritt aus den monomeren Anionen, vielfach in Kombination mit einer vorausgehenden Protonierung.

Polyborate kommen in der Natur vor oder können synthetisch im System Metalloxid/Boroxid/Wasser isoliert werden. Die meisten von ihnen sind wasserhaltig, besitzen also die Formel m Metalloxid $\cdot$ n Boroxid $\cdot x$ H_2O. Man kann aber auch wasserfreie Polyborate durch Schmelzen oder Erhitzen erhalten.

1.2. Zur Systematik

Gemessen an der bei den Phosphaten oder Silicaten erreichten tiefgehenden Systematik ist der Stand unserer Kenntnisse auf dem Gebiet der Borate unbefriedigend. Man ist zunächst geneigt, Parallelen zu den Silicaten und auch den Phosphaten zu suchen. Doch eine Analogie zwischen der Borsäure einerseits und der Kieselsäure bzw. Phosphorsäure andererseits besteht letztlich nur darin, daß alle in der Lage sind, Isopolysäuren zu bilden. Der erste wesentliche Unterschied liegt im Verhalten der *Borsäure* selbst, die im Gegensatz zur Monokieselsäure sowohl in freier Form als auch in Lösung unter normalen Bedingungen haltbar ist und nicht zur Kondensation neigt. In verdünnter wäßriger Lösung liegt sie monomer und größtenteils undissoziiert vor; ihre Dissoziationskonstante beträgt bei Konzentrationen $< 0,1$ m ca. 10^{-9}. Bei der Zugabe von OH$^-$-Ionen hydratisiert die B(OH)$_3$-Molekel zu einer Tetrahydroxoborsäure, die nach

$$B(OH)_3 + H_2O \rightleftharpoons B(OH)_3 \cdot H_2O \rightleftharpoons H[B(OH)_4] \rightleftharpoons H^+ + [B(OH)_4]^- \qquad [1,2]$$

unter Bildung von Tetrahydroxoborationen dissoziiert.

Aus dieser Tatsache ergibt sich ein wesentlicher Gesichtspunkt: Das Bor bestätigt gegenüber Sauerstoff außer der Koordinationszahl 3 auch die Koordinationszahl 4. Der Grund liegt im koordinativ ungesättigten Charakter der ebenen B(OH)$_3$-Molekel, der sie zu einer *Lewis-Säure* macht.

Die Beobachtung, daß die Acidität der Borsäure mit der Konzentration wächst, daß Polyborate Salze stärkerer Säuren als die der monomeren Borsäure sind, kann dadurch erklärt werden, daß kondensierte Strukturen mit vierbindigem Bor gebildet werden, die aus sterischen Gründen mit dreibindigem Bor nicht entstehen können. Beim Einengen der konzentrierten wäßrigen Lösung oder durch Verdrängung mit Hilfe starker Mineralsäuren läßt sich die Borsäure jedoch monomer zurückgewinnen. Die freie Säure kondensiert nicht; das wurde durch Titration bei verschiedenen Verdünnungen [3] festgestellt. Im Gegensatz zur freien Säure kristallisieren jedoch ihre Salze aus wäßriger Lösung mit Anionen, die durch Kondensation monomerer Teilchen entstanden sein müssen. Eine Ausnahme davon bilden nur die im stark alkalischen Bereich erhältlichen 1:1-Borate. In ihnen liegt, wie Ramanspektren [1] und Messungen mit der Bleiamalgamelelektrode [2] beweisen, nur das isolierte Ion [B(OH)$_4$]$^-$ vor.

1.3. Bindungsprinzipien in Boraten

Einige Strukturen wasserhaltiger Borate wurden aufgeklärt; aus den Ergebnissen wuchs ein Bindungsschema, das nur für Bor-Sauerstoff-Verbindungen charakteristisch ist. Einige Prinzipien der Bindung in Polyborat-Strukturen haben Edwards und Ross [4] postuliert:

1) In festen hydratisierten Boraten existiert Bor in drei- und vierfacher Koordination mit Sauerstoff (dreieckig-planar und tetraedrisch). Die Anzahl der vierbindigen Boratome entspricht der Ladung des Anions.

2) Die kristallisierten Polyborate sind monomer oder polymer. Die Existenz von Monomeren, Dimeren, Trimeren, Tetrameren, Pentameren und polydimensionalen Netzwerken wurde nachgewiesen.

3) In den Tri- und höheren Polyboraten ist die Grundstruktur ein sechsatomiger Ring mit alternierenden Bor- und Sauerstoffatomen; a) der Ring ist stabil, wenn er ein oder zwei vierbindige Boratome enthält; die mit drei vierbindigen Boratomen vorkommenden Verbindungen sind in wäßriger Lösung nicht beständig; b) diese Ringe können, unter Bildung eines zweiten Ringsystems, an den vierbindigen Boratomen kondensieren und höhere Polyanionen, z. B. Tetramere und Pentamere, bilden; c) die entstandenen Polyanionen können über Sauerstoffbrücken Ketten, Schichten und dreidimensionale Netzwerke bilden.

Ähnliche Regeln erarbeitete auch Dale [5] bei Arbeiten über Organo-Boratkomplexe. Ähnlich wie bei der Einteilung der Silicate unterscheidet nun Christ [6] noch zwischen Inselpolyborationen, bei denen die diskreten Einheiten nur über Wasserstoffbrückenbindungen im Kristall zusammengehalten werden, und zwischen Ketten- oder Schichtenpolyanionen, bei denen die sich wiederholenden Einheiten ein- oder zweidimensional durch Sauerstoffatome verbunden sind. Aus diesen Strukturforderungen stellte er Voraussagen für weitere mögliche Polyanionen auf.

Alle Strukturen auf dem Boratgebiet entsprechen diesen Postulaten mit Ausnahme des von Lehmann und Gaube [7] als $K_3[B_3O_3(OH)_6] \cdot x\,H_2O$ formulierten wasserhaltigen Kaliummetaborats und des von Clark [8] durch Röntgenstrukturanalyse und von Cuthbert u. Mitarb. [9] durch Einkristall-NMR-Spektren aufgeklärten Minerals Tunellit, $Sr[B_6O_9\text{-}(OH)_2] \cdot 3\,H_2O$.

1.4 Möglichkeiten zur Strukturaufklärung

Die Strukturen kristallisierter Borate durch Röntgenpulver- bzw. Röntgeneinkristallanalysen mit Strukturverfeinerung aufzuklären ist sehr schwierig, weil die Borate meist im triklinen oder monoklinen System kristallisieren.

Die Art der Koordination der Sauerstoffatome um das Boratom kann laut Silver und Bray [10] am besten durch ^{11}B-*NMR-Spektroskopie* ermittelt werden. Die Borate mit vierbindigem Bor spalten die ^{11}B-NMR-Resonanz scharf in eine „1. Ordnung" mit Quadrupol-Kopplungskonstanten, die gegen null gehen, während die mit dreibindigem Bor Übergänge „2. Ordnung" mit großen Kopplungskonstanten geben. Die Art der Bindung der H-Atome kann durch ^1H-NMR-Spektren festgestellt werden; befinden sie sich in OH-Gruppen mit einem H—H-Abstand von 2,3—2,7 Å, d.h. in Form von Strukturwasser, so erhält man im Breitlinienspektrum bei 90 °K eine schmale erste Ableitung der Protonenresonanzabsorptionslinie mit einem zweiten Moment $\Delta \bar{H}^2$ von < 15 Gauss^2, sind sie aber in H_2O mit einem H—H-Abstand von 1,5—1,6 Å, also in Form von Kristallwasser, enthalten, so ist die Linie breit und

$\Delta \bar{H}^2$ beträgt mehr als 15 Gauss[2]. Auch über eine eventuelle Wasserstoff-brückenbindung im Polyborat lassen sich aus der Größe des zweiten Moments Aussagen machen.

Die Anwendung von *Raman- und IR-Spektren* auf die Struktur-bestimmung von Polyboraten ist begrenzt. Wie Weir [11] mitteilt, lassen sich nur in wenigen einfachen Fällen die Schwingungsfrequenzen im IR-Bereich für bereits bekannte Strukturen berechnen und mit den Messungen vergleichen. Die Spektren für alle Borate, die neben vierbindigem auch dreibindiges Bor enthalten, sind für eine Strukturanalyse zu kompliziert. Dennoch sind oft Beziehungen zwischen den Spektren erkennbar, und es können gleiche Anionentypen identifiziert werden. Eine Unterscheidung zwischen drei- und vierbindigem Bor in wasserhaltigen Polyboraten ist möglich, aber weniger sicher als bei wasserfreien Boraten.

In den *Remissionsspektren* findet man bei kristallwasserhaltigen Verbindungen ausgeprägte H_2O-Banden. Weitere Verfahren zur eventuellen Unterscheidung zwischen Struktur- und Kristallwasser sind die *Thermogravimetrie* und der *isotherme Abbau*.

Eine Bestimmungsmöglichkeit für die Molekülstrukturen von Boraten in Lösung ist die Untersuchung von Phasengleichgewichten durch kryoskopische Messung und Aufstellung der Löslichkeitsisothermen. Daraus werden thermodynamische Größen wie Bildungswärmen, Lösungswärmen, Verdampfungswärmen, Schmelzwärmen und Entropien berechnet.

1.5. Strukturen von Polyboraten in Lösung

Die thermodynamischen Größen lassen auch Aussagen über Polyborate in wäßriger Lösung zu. Die Frage nach Existenz und Art von Polyborationen in Lösung wurde schon häufig zu klären versucht. In stark verdünnten Lösungen liegen nur monomere Teilchen, nämlich das saure $B(OH)_3$ und das alkalische $B(OH)_4^-$, nebeneinander vor. Löst man Borsäure $B(OH)_3$ und Metaborat BO_2^-, das zu Orthoborat $B(OH)_4^-$ hydratisiert, in verschiedenen Konzentrationen zusammen in Wasser auf, so erhält man bei Borkonzentrationen von $> 0,1$ molar Gleichgewichte zwischen verschiedenen Polyborationen, die sich unmeßbar schnell einstellen. Diese Polyborationen werden durch Kondensation bei der Reaktion zwischen $B(OH)_3$ und $B(OH)_4^-$ gebildet, z.B. nach:

$$2\,B(OH)_3 + B(OH)_4^- \rightleftharpoons [B_3O_3(OH)_4]^- + 3\,H_2O\,.$$

Da bei der Bildung der Polyborate Wasser frei wird, wird es verständlich, daß Polyborationen nur in konzentrierteren wäßrigen Lösungen existieren. Diese Polyboratbildung kann auf verschiedene Art und Weise bewiesen werden: durch Aufstellung der Löslichkeitsisothermen, wie

z. B. im System $Na_2O/ B_2O_3/ H_2O$ [12], im System $K_2O/ B_2O_3/ H_2O$ [13] oder im System $(NH_4)_2O/ B_2O_3/ H_2O$ [14], durch thermochemische Untersuchung der Wärmeentwicklung bei Zugabe von Borsäure zu Metaborat- oder Boraxlösung [15], durch Untersuchung der Verteilungsgleichgewichte in Amylalkohol [16], Octylalkohol [17,18] oder Isoamylalkohol [19,20], durch Ionenaustauschuntersuchungen [21], durch kryoskopische Untersuchungen [15,17,22], durch Messung der elektrischen Leitfähigkeit [23,24], durch Bestimmung der Wasserstoffionenkonzentration [25,26,27], durch IR-Spektren [28], durch Raman-Spektren [29] oder ^{11}B-NMR-Spektren [30].

Wie Ingri u. Mitarb. [31,32] zeigten, sind bei höheren Borkonzentrationen die Boratlösungen entweder saurer oder alkalischer, als für monomere Species zu berechnen ist.

Jede Titrationskurve, die der durchschnittlichen Ladung Z pro Boratom (oder der durchschnittlichen Zahl von OH^--Ionen, die an ein Boratom gebunden sind) in Beziehung zum p_H-Wert entspricht, geht durch einen gemeinsamen „isohydrischen" Punkt, der in 3 m $NaClO_4$-Lösung bei $Z = 0,4$ und $p_H = 9,00$ liegt. Carpeni [33] hatte aufgrund von EMK-Messungen diesen Punkt schon früher definiert und ihn als Maximum der Polyanionenkonzentration charakterisiert.

Während die Bildung von Polyionen in Boratlösungen bewiesen ist, ist über die Art der Polyionen bei den Autoren kaum Übereinstimmung vorhanden. Das Problem wird offensichtlich dadurch kompliziert, daß mehr als ein Polyanion anwesend ist, die unter bestimmten Bedingungen gebildeten Polyanionen nicht notwendigerweise auch unter anderen Bedingungen gebildet werden und daß die Bildungskonstanten im Gleichgewicht nicht groß sind. Die Deutung der Ergebnisse über mögliche Strukturen in Lösung werden durch schnelle Dissoziation der Polyborationen und Austauschbarkeit der Borkoordination erschwert. Raman-Spektren konzentrierter Boratlösungen sind erhältlich, doch sind die Frequenzen für die B—O-Schwingungen unscharf, und ihre Zuordnung ist nur schwer zu treffen*).

Für die Bestimmung der Bildungskonstanten der verschiedenen Polyborationen im Gleichgewicht, mit deren Hilfe Aussagen über Stöchiometrie und Strukturen in Lösung möglich wären, war eine neue Technik notwendig. Den Anfang machten einige Veröffentlichungen der letzten Jahre, die bewiesen, daß das Polyboration $[B_3O_3(OH)_4]^-$ mit seiner cyclischen Struktur das in Lösung vorherrschende Polyion ist. In einer neueren Arbeit hat Ingri [34] aufgrund der nach dem LETAGROP-System berechneten Gleichgewichtskonstanten die Verteilung des Bors auf die verschiedenen Ionen bei gegebener Konzentration und gegebenem p_H-Wert für 0,2 m Boratlösungen in Gegenwart von NaOH bestimmt. Da-

---

*) Persönliche Mitteilung von Prof. R. S. Tobias, University of Minnesota.

nach liegen an Polyionen neben $[B_3O_3(OH)_4]^-$ auch $[B_3O_3(OH)_5]^{2-}$, $[B_4O_5(OH)_4]^{2-}$ und $[B_5O_6(OH)_4]^-$ im Gleichgewicht vor. Aufgrund von *Messungen der chemischen Relaxationszeiten* mit Hilfe der Temperatursprungmethode nach Eigen kommen auch Anderson u. Mitarb. [35] dazu, die Bildungskonstanten im Gleichgewicht für die erwähnten Ionen zu berechnen. Neben diesen Polyionen werden von anderen Autoren auch andere Polyionen als in Lösung existent angesehen, nämlich von Lefebvre [36] einwertig negative tetramere Teilchen und von Carpéni [37] zweifach negative pentamere Teilchen.

Die Strukturen der Polyborationen in Lösung scheinen demnach weitgehend denen zu entsprechen, die in kristallisierten wasserhaltigen Polyboraten nachgewiesen wurden.

2. Experimentelles

Um die Strukturen einiger Kalium- und Natriumpolyborate ermitteln zu können, haben wir magnetische Protonenresonanzen gemessen. Dabei wurde versucht, widersprüchliche Literatur-Angaben zu klären. Auch wurde der thermische Abbau der Borate mit Hilfe von PMR-Messungen ausgewertet.

Außerdem haben wir zur Darstellung von Polyboraten ein erst vor wenigen Jahren [38] entwickeltes Verfahren herangezogen. Die Methode beruht auf der Hydrolyse von Borsäureester in Gegenwart von Basen in organischen Lösungsmitteln. Erfahrungsgemäß fallen die auf diesem Wege gewonnenen polysauren Salze kristallwasserfrei an, und somit erlaubt ihre Zusammensetzung direkte Rückschlüsse auf ihre Struktur. Es wurden auf diese Weise eine Anzahl neuer Polyborate erhalten und diese mit Hilfe von IR- und Remissionsspektren sowie durch ihre Röntgendaten charakterisiert.

Die Hydrolyse eines in einem *organischen Lösungsmittel* monomer gelösten Metallsäureesters führt in der Regel zu einem hochkondensierten Hydroxid oder Oxidhydrat. Sie wird also von einem Kondensationsvorgang begleitet. Bei der Kondensation müssen zwangsläufig als Zwischenprodukte Isopolysäuren bzw. deren Ester auftreten. Die Untersuchung der Vorgänge [39,40] hatte gelehrt, daß es möglich ist, durch Zugabe von Basen die intermediär gebildeten Polyanionen in Form von Salzen abzufangen. Unter geeigneten Reaktionsbedingungen lassen sich so auch hydrolyseempfindliche Salze von Polysäuren rein darstellen. Wichtig ist dabei, daß Wasser nur in geringen Mengen angeboten wird und damit einzig als Reagenz für die Hydrolyse des Esters und nicht als Solvens wirkt. Die Reaktion

$$\text{Ester} + \text{Base} + \text{Wasser} \rightleftharpoons \text{Salz} + \text{Alkohol}$$

läuft im allgemeinen quantitativ ab, wenn Ester und Base gegenüber Wasser im Überschuß vorhanden sind und wenn man Lösungsmittel benutzt, in denen das gebildete Salz unlöslich ist. In Lösung liegen fast immer mehrere Arten von Polyionen im Gleichgewicht miteinander vor. Bei der Zugabe einer Base mit einem bestimmten Kation entscheidet die Löslichkeit der verschiedenen Salze, die das Kation mit den vorhandenen Polyanionen bilden kann, darüber, welches Salz auskristallisiert. Bei der Esterhydrolyse, bei der organisches Lösungsmittel als Solvens wirksam ist, liegen oft andere Löslichkeitsverhältnisse als in wäßriger Lösung vor. Es können daher auch im wäßrigen System unbekannte Salze gebildet werden.

Der *Kondensationsgrad* der Polyanionen läßt sich auf eine einfache Weise variieren. Er hängt im wesentlichen vom p_K-Wert der Base im Hydrolysegemisch ab. Die Verhältnisse liegen also ähnlich wie in wäßriger Lösung, wo der Kondensationsgrad eines Polyanions bei gegebener Konzentration eine Funktion der Wasserstoffionenkonzentration ist.

Die Esterhydrolysenmethode bietet einen weiteren Vorteil, der für die Strukturaufklärung von Polyionen von Bedeutung ist. Im Gegensatz zu den fast immer mit Kristallwasser zu erhaltenden polysauren Salzen sind die im organischen Lösungsmittel gefällten Salze erfahrungsgemäß kristallwasserfrei, da das Wasser nicht als Solvens wirkt. Allerdings enthalten sie oft anstelle des Wassers *organisches Lösungsmittel gebunden.* Daß nur kristallwasserfreie Salze bei der Esterhydrolyse entstehen, ist damit zu erklären, daß die geringe Wassermenge zwar ausreicht, um einen Teil des überschüssigen Esters zu hydrolysieren, nicht aber, um sich in Form von Kristallwasser in die Hohlräume des Polyanions einzulagern. Ist der Raumbedarf des organischen Lösungsmittels ausreichend, so geht dieses in die Zwischengitterplätze der Salze. Durch eine quantitative Analyse des Salzes oder durch Bestimmung des Wasserverbrauchs bei der Hydrolyse kann man den Wassergehalt im Salz bestimmen. Dieses Wasser kann demnach nur als Strukturwasser, d.h. in Form von OH-Gruppen, gebunden sein. Damit hat man eine weitere Möglichkeit, dem strukturellen Aufbau von Polyionen auf die Spur zu kommen, denn für eine Strukturaufklärung von Polyionen ist die Kenntnis des Strukturwassergehaltes unerläßlich.

Diese Methode der Esterhydrolyse bietet somit eine wertvolle Ergänzung anderer Untersuchungsmethoden über den Zustand der festen polysauren Salze. Da in den aus organischem Lösungsmittel auskristallisierenden Salzen in der Regel Polyionen mit den gleichen Kondensationsgraden vorliegen, wie sie bei den aus wäßriger Lösung gewonnenen Verbindungen auftreten, kann man auch Aussagen über die Strukturen der aus wäßriger Lösung kristallisierenden Salze von Polysäuren machen.

3. Ergebnisse und Diskussion

3.1. Ammoniumpolyborate

3.1.1. Bekannte Ammoniumpolyborate

Ammoniumpolyborate kommen als Mineralien in der Natur vor. Das Mineral Ammonioborit hat die Zusammensetzung $(NH_4)_2O \cdot 5 B_2O_3 \cdot 5^1/_3 H_2O$ *(1)* [43)] und das Mineral Lardellerit ist $(NH_4)_2O \cdot 5 B_2O_3 \cdot 4 H_2O$ *(2)* [44)]. Synthetisch wurden Ammoniumpolyborate bis vor kurzem nur aus wäßrigen Lösungen gewonnen. Rhombisches $(NH_4)_2O \cdot 5 B_2O_3 \cdot 8 H_2O$ *(3)* [45,46)] bildet sich bei langsamer Abkühlung einer heißen konzentrierten Lösung von Borsäure und Ammoniak im molaren Verhältnis 3,5:1. Dieses Salz ist mit der entspr. Kaliumverbindung isomorph. Nach Tolédano [47)] gibt es auch eine β-Form von *3*, die durch längeres Rühren einer gesättigten wäßrigen Lösung der α-Form erhalten wird. Die α-Form ist bei Raumtemperatur nämlich metastabil. Die β-Form geht beim Erhitzen langsam in $(NH_4)_2O \cdot 5 B_2O_3 \cdot 5,5 H_2O$ über. Im ternären System $(NH_4)_2O/B_2O_3/H_2O$ tritt oberhalb von 100 °C eine Verbindung der Formel $(NH_4)_2O \cdot 5 B_2O_3 \cdot 4 H_2O$ auf, deren Zusammensetzung der von natürlichem Lardellerit gleicht, doch unterscheiden sich Mineral und synthetische Verbindung in ihren Röntgendiagrammen. Beim Erhitzen auf 150 °C gibt *3* sechs Mole Wasser ab unter Bildung von $(NH_4)_2O \cdot 5 B_2O_3 \cdot 2 H_2O$ *(4)*. Tetragonales $(NH_4)_2O \cdot 2 B_2O_3 \cdot 4 H_2O$ *(5)* [48)] bildet sich bei langsamer Abkühlung einer warmen Lösung von Borsäure und Ammoniak im molaren Verhältnis 0,47:1. *5* besitzt einen hohen Ammoniakdampfdruck. Mischungen von *3* und *5* reagieren beim Erwärmen zu einem Kristallkuchen, der $(NH_4)_2O \cdot 4 B_2O_3 \cdot 6 H_2O$ *(6)* [49)] enthält.

6 kristallisiert auch oberhalb 31,5 °C aus Lösungen, die etwas mehr Ammoniak enthalten als dem molaren Verhältnis 1:4 entspricht. Im Gegensatz zu anderen Autoren fanden Tolédano und Awka [14)], daß *6* bis 0 °C stabil sein kann. Versuche von Menzel [49)], ein Ammonium(1:1)-Borat durch Kühlen einer Lösung von *5* in Anwesenheit eines hohen Ammoniakdruckes oder durch Behandeln einer Suspension von *5* mit NH_3-Gas oder aber durch Erhitzen von festem *5* und Kühlen auf −15 °C im NH_3-Gasstrom zu erhalten, schlugen fehl. Es wurden nur Produkte erhalten, in denen eines der vier H_2O-Moleküle in *5* durch ein NH_3-Molekül ersetzt ist, so daß die Zusammensetzung $(NH_4)_2O \cdot 2 B_2O_3 \cdot 3 H_2O \cdot NH_3$ erreicht wird. Das Röntgenpulverdiagramm dieser Verbindung gleicht dem von *5*. Wird trockenes Borax mit gasförmigem NH_3 behandelt, so geht etwa ein Mol Wasser verloren, aber es wird nur sehr wenig NH_3 absorbiert. Wirkt NH_3-Gas bei Raumtemperatur auf Borsäure ein [50)], so entsteht unter Erwärmen ein schmieriges Produkt, aus dem bei Abbruch der Reaktion vor einer vollständigen Auflösung der Borsäure und Extrahieren des nicht umgesetzten Anteils an Borsäure mit Aceton reines *3* zu isolieren ist.

Auf die Idee, ein *polares organisches Lösungsmittel* als Reaktionsmedium zu benutzen, kamen gleichzeitig mit uns Lehmann und Schmidt [51]. Sie gingen von freier Borsäure aus. Interessanterweise erhielten sie beim Einleiten von NH_3 in eine siedende gesättigte Lösung von Borsäure in Acetonitril bis zu einem NH_3:B-Verhältnis von 1:5 nach einigen Tagen Kristalle von *3*, in Dioxan dagegen oder beim Eingießen der Reaktionslösung in viel Äther das auch von uns durch Boresterhydrolyse gewonnene, bisher aber nur als Mineral bekannte *2*. Beim Sättigen der Ausgangslösung mit NH_3 scheiden sich Kristalle von *5* ab. In siedendem Dimethylformamid bildet sich, unabhängig vom NH_3:B-Verhältnis, immer $(NH_4)_2O \cdot 5 B_2O_3 \cdot 4 H_2O \cdot 2 DMF$. Die Löslichkeiten im System $(NH_4)_2O/B_2O_3/H_2O$ wurden von Sborgi und Ferri [48] zwischen $-1°$ und $+90 °C$ untersucht. Wasserfreie Ammoniumborate der Form $(NH_4)_2O \cdot B_2O_3$ waren noch unbekannt.

3.1.2. Durch Boresterhydrolyse dargestellte Ammoniumpolyborate

Führt man die Hydrolyse [41] einer verdünnten Lösung von Borsäuretrimethyl- oder Borsäure-tri-n-propylester in Äther, Äthanol, Dioxan oder Testbenzin in Gegenwart von Ammoniak mit einem organischen Lösungsmittel durch, das 1 bis 10% Wasser enthält, so erhält man nach dem Trocknen die Polyborate

$$(NH_4)_2O \cdot 5 B_2O_3 \cdot 4 H_2O \; (2),$$

$$(NH_4)_2O \cdot 4 B_2O_3 \cdot 5 H_2O \; (7),$$

$$(NH_4)_2O \cdot 3 B_2O_3 \cdot 4 H_2O \; (8)$$

oder Gemische dieser Salze. Die Bedingungen, unter denen reine Verbindungen erhalten worden sind, hat Heller [41] angegeben. Welches Polyborat sich bildet, hängt von den Konzentrationsverhältnissen an Borsäureester, Ammoniak und Wasser ab. Aus Lösungen mit höherem Wassergehalt oder bei niedrigerer Boresterkonzentration erhält man Polyborate mit niedrigerem Kondensationsgrad. Die drei genannten Salze sind nach Röntgen-Diffraktionsaufnahmen kristallin und lösen sich leicht in Wasser.

Ein weiteres Ammoniumpolyborat wurde bei der raschen Zugabe von konzentrierter wäßriger Ammoniaklösung zu einer überschüssigen konzentrierten Lösung des Borsäuretrimethylesters in Dioxan in Form einer wasserfreien, röntgenamorphen, gummiartigen Substanz erhalten, die nach dem Trocknen ein weißes Pulver der Zusammensetzung $(NH_4)_2O \cdot 3 B_2O_3$ *(9)* ist, das sich in Wasser recht schwer löst.

Da nach den bisherigen Ergebnissen bei der Hydrolyse von Estern unter den genannten Bedingungen nur kristallwasserfreie Verbindungen erhalten wurden, ist anzunehmen, daß das bei der Analyse der neu dargestellten Ammoniumpolyborate gefundene H_2O jeweils als Strukturwasser gebunden ist. Die Kenntnis des Strukturwassergehaltes von *2, 7* und *8* erlaubt es uns nun, die Strukturformeln ihrer Anionen aufzustellen. Dabei muß man jedoch eines bedenken: analytisch läßt sich für jedes Salz nur eine Mindestformulierung ermitteln. Prinzipiell könnte es sich z. B. bei *2* statt um ein Pentaborat um ein Dekaborat oder um ein noch höher aggregiertes Teilchen handeln. Leider sind Molekulargewichtsbestimmungen unmöglich, da Polyborate beim Lösen in niedrigermolekulare Teilchen aufgespalten werden.

Das durch Esterhydrolyse gewonnene Salz *2* [41] entspricht in seiner Zusammensetzung dem Mineral Lardellerit und der bei über 100 °C dargestellten synthetischen Verbindung; es besitzt 4 Mole H_2O pro Formel $(NH_4)_2 \cdot 5 B_2O_3 \cdot x H_2O$ weniger Wasser als das aus wäßriger Lösung kristallisierende bekannte Salz *3*. Dieses aber ist isomorph mit dem Salz $K_2O \cdot 5 B_2O_3 \cdot 8 H_2O$, dessen Struktur durch Röntgenanalyse [52] und durch Protonenresonanzspektren [53,54] aufgeklärt wurde. Danach handelt es sich um ein Inselpolyborat, dem die Strukturformel $K(H_2O)_2 [B_5O_6(OH)_4]$ zuerteilt werden muß. Entsprechend müssen die Strukturformeln von *3* $NH_4(H_2O)_2[B_5O_6(OH)_4]$ und von *2* $NH_4[B_5O_6(OH)_4]$ lauten; *2* kann also wirklich ohne Kristallwasser formuliert werden. Auch im Mineral Ammonioborit *(1)* und in $(NH_4)_2O \cdot 5 B_2O_3 \cdot 4 H_2O \cdot 2 DMF$ liegt wahrscheinlich das gleiche Pentaboration wie in *2* oder *3* vor. Das Abbauprodukt *4* kann jedoch kein Inselpolyborat sein, dazu enthält es zu wenig Wasser. Unter H_2O-Austritt haben jeweils 2 OH-Gruppen reagiert, und die durch O-Brücken verbundenen Struktureinheiten bilden Ketten oder Schichten mit dem Anion $\{B_5O_7(OH)_2\}_n^{n-}$.

Von den Ammoniumtetraboraten scheint *5* das Anion des Borax $[B_4O_5(OH)_4]^{2-}$, zu besitzen; dieses Anion wurde bei der Kristallstrukturanalyse des analogen Kaliumsalzes von Marezio *et al.* [55] gefunden. Das aus wäßriger Lösung gewonnene *6* enthält in der Formel $(NH_4)_2O \cdot 2 B_2O_3 \cdot x H_2O$ 1 Mol Wasser mehr als das durch Boresterhydrolyse erhaltene *7*. Dale [5] hält es für möglich, daß das Anion des von Filsinger [56] beschriebenen $Li_2O \cdot 2 B_2O_3 \cdot 10 H_2O$ die Formel $[B_4O_4(OH)_5]^-$ hat. Wenn man diese Anionenformel für *7* einsetzt, erhält man $NH_4 [B_4O_4(OH)_5]$ und damit eine kristallwasserfreie Struktur. Bei *6* handelt es sich danach wahrscheinlich um $(NH_4)_2(H_2O) [B_4O_4(OH)_5]_2$. Auf die Existenz des Anions $[B_4O_7H.aq]^-$ in wäßriger Lösung wurde bereits von Lefebvre [36] bei pH-Messungen, von Thygesen [23] aufgrund von Leitfähigkeitsuntersuchungen und von Everest und Popiel [21] bei Ionen-austauschuntersuchungen geschlossen; schreibt man aq $= 2\,H_2O$, so erhält man das in den Kristallen vermutete $[B_4O_4(OH)_5]^-$-Ion.

Ein Ammoniumtriborat war bisher unbekannt, allerdings kennt man die Strukturen von Triboraten mit Erdalkalimetallionen sehr gut. Bei dem durch Esterhydrolyse gewonnenen Salz *8* handelt es sich wohl um $NH_4[B_3O_3(OH)_4]$; das Ion $[B_3O_3(OH)_4]^-$ ist sowohl von Ingri [57] bei Gleichgewichtsuntersuchungen in Lösung als auch von Anderson *et al.* [35] bei Messungen der chemischen Relaxationszeiten als beherrschende Gleichgewichtskomponente bei p_H9 im System der Natriumpolyborat-lösungen erkannt worden.

Die röntgenamorphe Verbindung $(NH_4)_2O \cdot 3\,B_2O_3$ kann nicht mono-mer sein; sie soll daher als polymeres Ammoniumtriborat $\{NH_4B_3O_5\}_n$ formuliert werden. Die Röntgenstrukturanalyse der analogen Verbindung $Cs_2O \cdot 3\,B_2O_3$ durch Krogh-Moe [58] erbrachte den Beweis eines polymeren Anions $\{B_3O_5\}_n^{n-}$, das durch Verbindung von sechsatomigen Ringen mit alternierenden Bor- und Sauerstoffatomen über Sauerstoffbrücken zu Helices entsteht, die dann, miteinander vernetzt, ein dreidimensionales Gerüst bilden. Diese Struktur dürfte auch in dem von uns dargestellten Ammoniumpolytriborat vorliegen.

Betrachtet man die bisher existierenden Ammoniumpolyborate, so fällt auf, daß in ihnen mit einer Ausnahme nur einfach negativ geladene Polyborationen enthalten sind. Diese eine Ausnahme, nämlich das aus konzentrierten wäßrigen Lösungen langsam auskristallisierende oder sich durch Einwirkung von NH_3-Gas auf feuchte Kristalle von *3* um-wandelnde $(NH_4)_2[B_4O_5(OH)_4]$, besitzt aber einen hohen Ammoniak-dampfdruck und gibt daher leicht Ammoniak ab. Die relativ schwache Base Ammoniak kann also mit Polyborsäuren nur relativ basearme Salze bilden.

3.2. Lithiumborate

Im System $Li_2O/B_2O_3/H_2O$ existieren, wie Löslichkeitsversuche zwi-schen 10 °C und 80 °C [59] sowie 30 °C und 100 °C [60] zeigen, die Verbin-dungen

$$Li_2O \cdot B_2O_3 \cdot 16\,H_2O \; (1),$$

$$Li_2O \cdot 2\,B_2O_3 \cdot 4\,H_2O \; (2),$$

$$Li_2O \cdot 2\,B_2O_3 \cdot 3\,H_2O \; (3) \; und$$

$$Li_2O \cdot 5\,B_2O_3 \cdot 10\,H_2O \; (4).$$

Aufgrund des thermischen Abbaus und von Röntgenstrukturunter-suchungen kann *1* als $Li(H_2O)_6[B(OH)_4]$ geschrieben werden, denn es geht bei 40 °C in $Li[B(OH)_4]$ über, dessen Struktur [61] bestimmt wurde. $Li[B(OH)_4]$ wird bei 150 °C zu $\{Li[BO(OH)_2]\}_n$ und bei 300 °C zur γ-Modifikation von $LiBO_2$ umgewandelt. Wenn man *1* unter Atmosphärendruck auf über 200 °C erhitzt [62], so bildet sich zunächst γ-$LiBO_2$, das bei 350 °C in β-$LiBO_2$ und bei 580 °C in α-$LiBO_2$ umgewandelt wird. In α-$LiBO_2$ [63] sind die Boratome dreibindig, die BO_3-Dreiecke hängen mit je einer Ecke zusammen und bilden Ketten um eine zweizählige Schraubenachse. Dagegen sind in β-$LiBO_2$ [62] die Boratome vierbindig, und es bildet sich Tiefcristobalit-Raumnetzstruktur aus. In γ-$LiBO_2$ schließlich finden wir eine Struktur mit dreibindigem Bor, die der Doppelkettenstruktur des kubischen B_2O_3 oder der von CaB_2O_4 gleicht. Von den Lithiumtetraboraten ist *2* nur in Berührung mit seiner gesättigten Lösung beständig; zu isolieren ist nur *3*. Da beider Abbau bei 260 °C zu $Li_2B_4O_7$ führt, dessen Struktur aufgeklärt [64] wurde, kann man II als $Li_2(H_2O)_2$ $[B_4O_5(OH)_4]$ und *3* als $Li_2(H_2O)[B_4O_5(OH)_4]$ formulieren. Das nur bei Temperaturen über 40 °C kongruent lösliche *4* zersetzt sich bei 130 °C zu $Li_2O \cdot 5\,B_2O_3 \cdot 2\,H_2O$, dann zum Monohydrat und schließlich bei 400 °C zu LiB_5O_8; *4* kann wohl als $Li(H_2O)_3[B_5O_6(OH)_4]$ formuliert werden. Das früher [56] beschriebene $Li_2O \cdot 4\,B_2O_3 \cdot 10\,H_2O$ kommt in modernen Veröffentlichungen nicht mehr vor; in dieser Verbindung hatte Dale [5] das Anion $[B_4O_4(OH)_5]^-$ vermutet.

Wasserfrei gibt es Lithiumborate mit den $Li_2O:B_2O_3$-Verhältnissen 6:1, 3:1, 3:2, 2:1, 1:1 (drei Modifikationen), 2:4, 2:5, 1:3, 1:4 und 1:5. Ramanspektren von Schmelzen in diesem System zeigen Ähnlichkeiten mit denen von Polyboraten bekannter Struktur [65].

3.3. Natriumborate

Im System $Na_2O/B_2O_3/H_2O$ existieren als Bodenkörper

$$Na_2O \cdot B_2O_3 \cdot 8\,H_2O \; (1),$$

$$Na_2O \cdot B_2O_3 \cdot 4\,H_2O \; (2),$$

$$Na_2O \cdot 2\,B_2O_3 \cdot 10\,H_2O \; (3),$$

$$Na_2O \cdot 2\,B_2O_3 \cdot 5\,H_2O \; (4),$$

$$Na_2O \cdot 2\,B_2O_3 \cdot 4\,H_2O \; (5),$$

$$2\,Na_2O \cdot 5\,B_2O_3 \cdot 7\,H_2O \; (6),$$

$$Na_2O \cdot 5\,B_2O_3 \cdot 10\,H_2O \; (7) \; \text{und}$$

$$Na_2O \cdot 5\,B_2O_3 \cdot 2\,H_2O \; (8).$$

Einige kommen als Mineralien vor, nämlich *3* als Borax oder Tincal, *4* als Tincalconit oder Mohavit, *5* als Kernit oder Rasorit, *6* als Ezcurrit und 7 als Sborgit. Eine weitere, nicht in diesem System als Bodenkörper vorkommende Verbindung ist das Mineral Nasinit,

$$2 \text{ Na}_2\text{O} \cdot 5 \text{ B}_2\text{O}_3 \cdot 5 \text{ H}_2\text{O} \ (9),$$

das auch bei 115—120 °C im geschlossenen Rohr synthetisch darzustellen ist [66]. Alle diese Verbindungen sowie ihre vermutlichen oder bekannten Strukturformeln sind in Tabelle 1 aufgeführt.

Tabelle 1. *Wasserhaltige Natriumborate*

Nr.	$\text{Na}_2\text{O}:\text{B}_2\text{O}_3:\text{H}_2\text{O}$	Vorkommen oder Darst. aus	Strukturformel
1	1:1:8	alkal. Borax-Lsg.	$\text{Na}(\text{H}_2\text{O})_2[\text{B}(\text{OH})_4]$ *
2	1:1:4	1 im Vakuum	$\text{Na}[\text{B}(\text{OH})_4]$?
3	1:2:10	Mineral:Borax	$[\text{Na}(\text{H}_2\text{O})_4]_2[\text{B}_4\text{O}_5(\text{OH})_4]$ *
4	1:2:5	Mineral:Tinkalkonit	$[\text{Na}_2(\text{H}_2\text{O})_3] \ [\text{B}_4\text{O}_5(\text{OH})_4]$
5	1:2:4	Mineral:Kernit	$\{[\text{Na}_2(\text{H}_2\text{O})_3] \ [\text{B}_4\text{O}_6(\text{OH})_2]\}_n$ *
10	1:2:2	Therm. Abbau v. 4	$\text{Na}_2[\text{B}_4\text{O}_5(\text{OH})_4]$
6	2:5:7	Mineral:Ezcurrit	$[\text{Na}_2(\text{H}_2\text{O})] \ [\text{B}_5\text{O}_6(\text{OH})_5]$?
9	2:5:5	Mineral:Nasinit	$\{[\text{Na}_2(\text{H}_2\text{O})] \ [\text{B}_5\text{O}_7(\text{OH})_3]\}_n$?
7	1:5:10	Mineral:Sborgit	$[\text{Na}(\text{H}_2\text{O})_3] \ [\text{B}_5\text{O}_6(\text{OH})_4]$

* Röntgenstrukturanalyse

3.3.1. Natriummonoborate

Das Monoborat 1 läßt sich durch Lösen von einem Mol Borax in warmer carbonatfreier 0,5 n NaOH-Lösung unter Ausschluß von CO_2 und Abkühlenlassen der übersättigten Lösung auf Raumtemperatur im H_2SO_4-Exsikkator darstellen [67]. 1 enthält nach seiner Röntgenstrukturanalyse [68] diskrete $\text{B}(\text{OH})_4^-$-Gruppen mit tetraedrisch koordinierten B-Atomen und oktaedrisch koordinierten Na^+-Ionen, seine Struktur kann durch die Formel $\text{Na}(\text{H}_2\text{O})_2[\text{B}(\text{OH})_4]$ wiedergegeben werden.

2 kann [67] durch Kühlen noch stärker alkalischer, auf 100 °C erhitzter Boraxlösungen, durch Erhitzen einer Aufschlämmung von 1 auf Temperaturen über den Umwandlungspunkt 53,6 °C und äußerst langsames Herunterkühlen oder durch Entwässern im Vakuumexsikkator über 80%iger H_2SO_4 erhalten werden. Die Kristalldaten von 2 sind bekannt [69], aber seine Struktur ist noch nicht bestimmt. 2 wandelt sich in Gegenwart von Wasser wieder leicht in 1 um. Diese reversible Umwandlung von 1 läßt die Struktur $\text{Na}[\text{B}(\text{OH})_4]$ vermuten.

Für die Entwässerung von *2* gibt es [67,69,70)] widersprüchliche Angaben, besonders hinsichtlich der Phasenübergangstemperaturen. Nach allen bisherigen Untersuchungen scheinen ein amorphes $Na_2O \cdot B_2O_3 \cdot 2 H_2O$ und ein kristallines $Na_2O \cdot B_2O_3 \cdot H_2O$ Zwischenprodukte beim Abbau zu $Na_2O \cdot B_2O_3$ zu sein.

3.3.2. Natriumtetraborate

Die Struktur von Borax *(3)* ist von Morimoto [71)] aufgeklärt worden. Ketten der oktaedrisch umgebenen Na^+-Ionen bauen über isolierte, nur durch Wasserstoffbrückenbindungen verbundene $[B_4O_5(OH)_4]^{2-}$-Gruppen Schichten auf. Das Anion enthält zwei dreibindige Boratome, die von Sauerstoff in einem Abstand zwischen 1,32 und 1,40 Å umgeben sind, und zwei vierbindige Boratome, deren Abstände zum nächsten Sauerstoffatom zwischen 1,46 und 1,54 Å betragen. Die Strukturformel von *3* kann am besten mit $[Na(H_2O)_4]_2[B_4O_5(OH)_4]$ wiedergegeben werden. Diese Formel bestätigt auch die Aufnahme des Protonenresonanzspektrums [72)]. Dabei beweist die erste Ableitung der Protonenresonanzabsorptionskurve bei 90 °K (Abb. 1a) sowohl das Vorhandensein von Kristallwasser als auch das von Strukturwasser. Das relativ große zweite Moment $\Delta \bar{H}^2$ von 25,1 Gauss^2 deutet darauf hin, daß sehr viel Kristallwasser neben wenig Strukturwasser vorliegt.

Tinkalkonit *(4)* läßt sich durch 100 h thermischer Entwässerung von Borax bei 60 °C darstellen. Christ und Garrels [73)] schlagen die Formel $[Na_2(H_2O)_3][B_4O_5(OH)_4]$ vor, die auch den Ergebnissen von ^{11}B-NMR-Untersuchungen [74,75)] entspricht. Dagegen schließen Dharmatti *et al.* [76)] aus der Linienform des bei 298 °K aufgenommenen Protonenresonanzspektrums auf die Formel $\{[Na(H_2O)_2]_2[B_4O_6(OH)_2]\}_n$. Aufgrund des von uns [72)] bei 90 °K aufgenommenen Protonenresonanzspektrums,

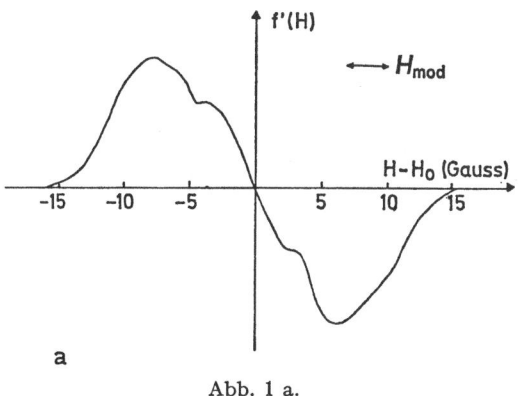

a

Abb. 1 a.

dessen erste Ableitung Abb. 1b zeigt, erscheint aber der Anteil der Hydroxidprotonen an der Gesamtzahl der Protonen in *4* größer zu sein, als es letztere Formel angibt: erstens ist $\Delta \bar{H}^2$ mit 21,5 Gauss^2 für derartig stark kristallwasserhaltige Verbindungen zu klein, zweitens spricht das Intensitätsverhältnis zwischen äußerem und innerem Maximum der Resonanzkurve gegen ein 4:1-Verhältnis von Kristallwasser- und Strukturwasserprotonen.

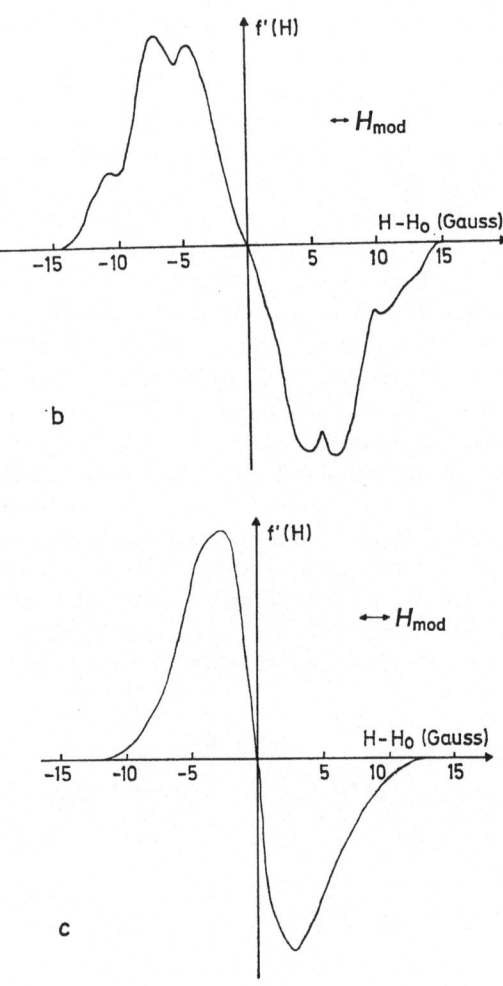

Abb. 1 a—c. Erste Ableitungen der Protonenresonanzabsorptionskurven bei 90 °K

a) $Na_2O \cdot 2\ B_2O_3 \cdot 10\ H_2O$; $\Delta \bar{H}^2 = 25,1$ Gauss^2;

b) $Na_2O \cdot 2\ B_2O_3 \cdot 5\ H_2O$; $\Delta \bar{H}^2 = 21,5$ Gauss^2;

c) $Na_2O \cdot 2\ B_2O_3 \cdot 2\ H_2O$; $\Delta \bar{H}^2 = 9,9$ Gauss^2

Weiterhin spricht der thermische Abbau von *4* in 50 h bei 85 °C zu $Na_2O \cdot 2 B_2O_3 \cdot 2 H_2O$ *(10)*, das auch Abbauprodukt von Borax ist, für eine Strukturformel mit dem Ion $[B_4O_5(OH)_4]^{2-}$. Die Linienform der ersten Ableitung des Protonenresonanzspektrums von *10* (Abb. 1c) und der Wert von $\Delta \bar{H}^2$ mit 9,9 Gauss^2 zeigen eindeutig, daß in *10* nur noch Strukturwasser vorliegt und ihm die Formel $Na_2[B_4O_5(OH)_4]$ zukommt. Wäre $\{[Na(H_2O)_2]_2[B_4O_6(OH)_2]\}_n$ für *4* richtig, dann sollte man in *10* nach Abbau von drei Molen Wasser noch Kristallwasser nachweisen können.

Nach Dharmatti hätte *4* auch das gleiche Polyion, das im Kernit *(5)* vorliegt, und somit sollte ein dem *5* entsprechendes 4-Hydrat beim Entwässern von *4* auftreten. Tatsächlich läßt sich *4* reversibel zu *3* anwässern [82].

Synthetisch [77] kann man *5* erst durch mehrtägiges Erhitzen von entwässertem, reichlich 4,5 Mol enthaltenden *4* im verschlossenen Rohr auf 130—135 °C, allmähliches Herunterkühlen auf 110 °C und Abschrecken erhalten. Nach Christ und Garrels [73] sollte die Umwandlung von Borax in Kernit bei einem Tiefendruck von 650—900 m unter der Erdoberfläche schon bei 53—63 °C eintreten. Aufgrund von ^{11}B-NMR-Untersuchungen [75] wurde für Kernit die Kettenstruktur $\{[Na_2(H_2O)_3][B_4O_6(OH)_2]\}_n$ vorgeschlagen. Nach Ross und Edwards [78] kann man sich das Anion $\{B_4O_6(OH_2)_n^{2n-}$ durch Kondensation von $[B_4O_5(OH)_4]^{2-}$-Ringen des Borax entstanden denken. Genau das fanden bei den Kristallstrukturanalysen Giese [79] sowie Cialdi u. Mitarb. [80]: die sechsgliedrigen Ringe werden über gemeinsame vierbindige Boratome so verkettet, daß sich spiralige Ketten in Richtung der kristallographischen b-Achse bilden. In den Ketten stehen die Ebenen der Ringe jeweils nahezu rechtwinklig aufeinander. In jeder Elementarzelle befinden sich zwei Ketten. Diese Struktur erklärt die Faserigkeit und Spaltbarkeit des Kernits sowie die Schwierigkeit seiner Darstellung aus Borax, denn dabei müssen B—O-Bindungen aufgebrochen werden.

Während *3* oder *4* beim Abbau [77] amorphes *10* liefern, bildet sich aus *5* ein kristallisiertes Produkt gleicher Zusammensetzung wie *10*, das aber vermutlich mit Kristallwasser als $\{[Na_2(H_2O)][B_4O_6(OH)_2]\}_n$ zu formulieren ist. Ein stark hygroskopisches Monohydrat, $Na_2O \cdot 2 B_2O_3 \cdot H_2O$ [81], wird durch Eindampfen von Boraxlösungen zur Trockne unter Druck oberhalb 177 °C dargestellt.

3.3.3. Natriumpentaborate

Von den Pentaboraten werden dem Ezcurrit *(6)* [82], $2 Na_2O \cdot 5 B_2O_3 \cdot 7 H_2O$, das Ion $[B_5O_6(OH)_5]^{2-}$ zugeschrieben. *6* kristallisiert langsam aus Lösungen aus, die an *7* und *3* bei 40 °C gesättigt und in bezug auf *6* übersättigt sind.

Nicht gesichert ist auch, ob das Kettenpolyboration $\{B_5O_7(OH)_3\}_n^{2n-}$ im Nasinit *(9)* vorliegt, das auch Augers Natriumborat [66] heißt.

Sborgit *(7)* [83] hat die gleichen Röntgenbeugungsbilder wie das synthetisch durch Kristallisation von *3* und Borsäure im $Na_2O:B_2O_3$-Verhältnis 1:5 dargestellte $Na_2O \cdot 5\,B_2O_3 \cdot 10\,H_2O$. Die Röntgendaten sind bekannt [84], aber nicht die Struktur. *7* enthält wahrscheinlich das gleiche Ion $[B_5O_6(OH)_4]^-$ wie das durch eine Röntgenstrukturanalyse bekannte synthetische $K_2O \cdot 5\,B_2O_3 \cdot 8\,H_2O$ [52]. Beim Erhitzen gibt *7* rasch Wasser ab unter Bildung des schwer zu isolierenden $Na_2O \cdot 5\,B_2O_3 \cdot 4\,H_2O$ [12]; bei 100 °C wird es zu $Na_2O \cdot 5\,B_2O_3 \cdot 2\,H_2O$ [85].

Durch achttägiges Rühren einer Aufschlämmung von *7* in wenig Wasser bei 94 °C erhielten Nies und Hulbert [12] eine Verbindung der Zusammensetzung $2\,Na_2O \cdot 9\,B_2O_3 \cdot 11\,H_2O$, die nur unter ihrer eigenen Lösung bei höheren Temperaturen als 57 °C beständig ist.

3.3.4. Wasserfreie Natriumborate

Die Existenz der wasserfreien Natriumborate mit den $Na_2O:B_2O_3$-Verhältnissen 1:1, 1:2, 1:3 und 1:4 hat schon Ponomareff [86] beschrieben. Morey und Merwin [87] berichten über eine 2:1-Verbindung, obwohl eine solche im Phasendiagramm nicht erscheint.

Die Struktur des Natriummetaborats, $Na_2O \cdot B_2O_3$, besteht nach Zachariasen u.a. [88] aus planaren $B_3O_6^{3-}$-Ringen, die durch die äußeren drei Sauerstoffatome miteinander verbunden sind. Dabei beträgt die Bindungslänge jedes Boratoms zu einem Sauerstoffatom 1,28 Å, zu zwei anderen Sauerstoffatomen 1,43 Å.

Das Natriumtetraborat $Na_2O \cdot 4\,B_2O_3$ enthält zwei unendlich lange, miteinander vernetzten B—O-Ketten [89], die sowohl Einzel- als auch Doppelringe enthalten. Die Natriumionen verbinden diese Ketten zu einer Bandstruktur.

3.3.5. Natriumcalciumborate

Die Minerale

$$\text{Ulexit, } Na_2O \cdot 2\,CaO \cdot 5\,B_2O_3 \cdot 16\,H_2O \quad \text{(1),} \quad \text{und}$$

$$\text{Probertit, } Na_2O \cdot 2\,CaO \cdot 5\,B_2O_3 \cdot 10\,H_2O \quad \text{(2),}$$

sind bekannt. Sowohl das trikline *1* als auch das monokline *2* wurden röntgenographisch untersucht [90] und die Gittergrößen angegeben. Eine neue Strukturanalyse von *2* [91] ergab, daß das dreifach negativ geladene Pentaboration in Form von Ketten der Zusammensetzung $\{B_5O_7(OH)_4\}_n^{3n-}$ vorliegt, die sich entlang der röntgenographischen c-Achse des Kristalls erstrecken. Mithin ist *2* als $\{[NaCa(H_2O)_3][B_5O_7(OH)_4]\}_n$ zu formulieren.

Laut Kristallstrukturanalyse enthält *1* [92)] isolierte Inselpolyborationen der Form $[B_5O_6(OH)_6]^{3-}$. Es muß $[NaCa(H_2O)_5][B_5O_6(OH)_6]$ geschrieben werden. *1* kann synthetisch auch in Form kleiner Nadeln aus Lösungen gewonnen werden, die Ca^{2+}-Ionen, Borax und Na^+-Ionen im Verhältnis $B_2O_3:Na_2O = 1,7:1$ enthalten. *2* kann man nur durch 4 Wochen langes Erwärmen von angefeuchtetem *1* auf 80—100 °C erhalten, wenn man außerdem die Lösung mit Kristallen von *2* impft.

3.4. Kaliumborate

In der Natur werden nur Mineralien gefunden, in denen Kalium neben anderen Kationen an Polyborationen gebunden ist. Z.B. findet man in Staßfurt das Mineral Kaliborit der Zusammensetzung $K_2O \cdot 4\,MgO \cdot 11\,B_2O_3 \cdot 18\,H_2O$.

Das Phasengleichgewichtssystem $K_2O/B_2O_3/H_2O$ wurde immer wieder untersucht, zuletzt von Carpéni [37)]. Eine Zusammenstellung der bekannten wasserhaltigen Kaliumborate enthält Tabelle 2.

3.4.1. Kaliummonoborate

Eine Untersuchung des Phasengleichgewichtes im Bereich der Borate mit dem $K_2O:B_2O_3$-Verhältnis $1:1$ zeigt, daß die feste Phase unterhalb 24 °C nur $K_2O \cdot B_2O_3 \cdot 8\,H_2O$ *(1)* ist. Diese Verbindung kristallisiert auch bei mehrstündigem Rühren einer Lösung dieser Zusammensetzung bei -60 °C aus; sie ist wahrscheinlich ein Orthoborat mit dem Anion $[B(OH)_4]^-$.

Zwischen 24 °C und 195 °C besteht die feste Phase aus einem wasserhaltigen Kaliummetaborat *(2)*. Über den Wassergehalt und die Struktur bestehen Meinungsverschiedenheit [93,96)]. Nachdem früher von Atterberg [45)] angenommen wurde, es handele sich um $K_2O \cdot B_2O_3 \cdot 3\,H_2O$, etwas später von Dukelski [95)], es sei $K_2O \cdot B_2O_3 \cdot 2,5\,H_2O$, bestimmten Lehmann und Gaube [7)] die Summenformel von *2* zu $K_2O \cdot B_2O_3 \cdot 2,67\,H_2O$ aufgrund des thermischen Abbaus zwischen 100 °C und 120 °C zu $K_2O \cdot B_2O_3 \cdot 0,67\,H_2O$. Nach kalorischen Messungen von Rollet und Tolédano [96)] soll *2* jedoch 2,5 Mole Wasser und das Produkt des thermischen Abbaus 0,5 Mol Wasser enthalten. Das bestätigte Tolédano [94)] später noch einmal durch Untersuchungen der Löslichkeit von *2* ohne und unter Druck, der thermischen Analyse und der chemischen Analyse des isolierten Produktes. Zviedre u. Mitarb. [97)] formulieren in jüngster Zeit die Verbindung wieder als $K_2O \cdot B_2O_3 \cdot 2,67\,H_2O$; diese Formulierung soll hier benutzt werden.

Zwischen 195 °C und 250 °C liegt als Gleichgewichtsphase danach $K_2O \cdot B_2O_3 \cdot 0,67\,H_2O$ *(3)* vor, oberhalb 250 °C das wasserfreie $K_2O \cdot B_2O_3$.

Tabelle 2. *Wasserhaltige Kaliumborate*

Nr.	$K_2O:B_2O_3:H_2O$	Vorkommen od. Darst. aus	Strukturformel (*Röntgenstrukturanalyse)
1	1:1:8	wäßr. Lsg. v. H_3BO_3 in KOH (1:1)	$K(H_2O)_4[B(OH)_4]$?
2	1:1:2,67 oder 1:1:2,5	erwärmter Lsg. v. H_3BO_3 in KOH (1:1)	$\{[K(H_2O)]_3[B_3O_5(OH)_2]\}_n$
4	1:1:2	Partielle Anwässerg. v. *3*	?
3	1:1:0,67 oder 1:1:0,5	Therm. Abbau v. *2*, 150 °C	$\{K_3[B_3O_5(OH)_2]\}_n$
5	1:2:8	KOH u. H_3BO_3 (1:1,6) in wenig H_2O	?
6	1:2:4	K_2CO_3 u. H_3BO_3 (1:2,5) in H_2O	$[K(H_2O)]_2[B_4O_5(OH)_4]$*
7	1:2:2	Erhitzen v. *6* (100 °C) od. Esterhydrolyse	$K_2[B_4O_5(OH)_4]$
8	1:2:1	Erhitzen v. *6* (145 °C)	$\{K_2[B_4O_6(OH)_2]\}_n$
9	2:5:5	KOH u. B_2O_3 (1:2,5; 115 °C—120 °C) od. Esterhydrolyse	$K_2[B_5O_6(OH)_5]$
14	1:3:4	Esterhydrolyse	$K[B_3O_3(OH)_4]$
10	1:5:8	wäßr. Lsg. v. H_3BO_3 in KOH (3:1)	$K(H_2O)_2[B_5O_6(OH)_4]$*
12	1:5:4	Esterhydrolyse	$K[B_5O_6(OH)_4]$
11	1:5:2	Therm. Abbau v. *10* (80—100 °C)	$\{K[B_5O_7(OH)_2]\}_n$
13	1:12:7	Esterhydrolyse	$\{K[B_{12}O_{15}(OH)_7]\}_n$

2 kann durch Glühen von H_3BO_3 und K_2CO_3 im Molverhältnis 2:1, Lösen in verdünnter wäßriger KOH und Trocknen im Exsikkator über konz. H_2SO_4 oder aber aus heißen stark konzentrierten Lösungen von Kaliumhydroxid und Borsäure im K:B-Verhältnis 1,25:1 durch langsames Abkühlen, Umkristallisieren aus Wasser, Waschen mit Aceton und Äther sowie Trocknen bei 100 °C erhalten werden [7]. Beim thermischen Abbau gibt *2* zwischen 100 °C und 150 °C 2 Mole Wasser ab und geht in *3* über, das bei 170 °C zu kristallisiertem $K_2O \cdot B_2O_3$ wird. Bei dessen partieller Anwässerung entsteht $K_2O \cdot B_2O_3 \cdot 2 H_2O$ *(4)*, das beim Erhitzen auf 120 °C ohne Zwischenstufe wieder zu $K_2O \cdot B_2O_3$ abgebaut wird.

Die Struktur des hexagonal mit $a = 12{,}75$ und $c = 7{,}33$ Å kristallisierenden $K_2O \cdot B_2O_3$ hat Zachariasen [98] aufgeklärt. Im Gitter bildet sich ein sechsatomiger Ring mit alternierend drei Bor- und drei Sauerstoffatomen. An jedem Boratom ist außerdem ein weiteres Sauerstoffatom gebunden; alle drei Sauerstoffatome umgeben jedes Boratom nahezu in gleichem Abstand und Winkel; jedes Kaliumion hat sieben Sauerstoffatome als nächste Nachbarn. Neuere NMR-Untersuchungen ergaben indessen [99] Quadrupolkopplungskonstanten, die das Bor in vierfacher Koordination erscheinen lassen.

In Anlehnung an die aus der Kristallstrukturanalyse resultierende Strukturformel $\{K_3B_3O_6\}_n$ für das wasserfreie Metaborat formulierten Lehmann und Gaube [7] 2 als $K_3[B_3O_3(OH)_6] \cdot H_2O$, 4 als $K_3[B_3O_3(OH)_6]$ und 3 als $\{K_3[B_3O_5(OH)_2]\}_n$.

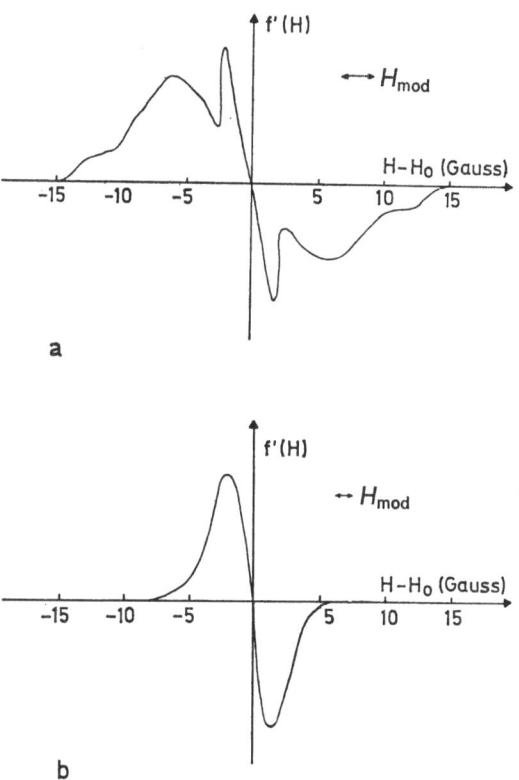

Abb. 2 a u. b. Erste Ableitungen der Protonenresonanzabsorptionskurven bei 90 °K

a) $K_2O \cdot B_2O_3 \cdot 2{,}67\ H_2O$; $\Delta \bar{H}^2 = 21{,}0\ \text{Gauss}^2$;

b) $K_2O \cdot B_2O_3 \cdot 0{,}67\ H_2O$; $\Delta \bar{H}^2 = 3{,}6\ \text{Gauss}^2$

Die Strukturvorschläge wurden von uns [72] mit Hilfe der Protonen-resonanzmethode bei 90 °K überprüft. Wie Abb. 2a zeigt, wurde für *2* in der ersten Ableitung der Absorptionskurve eine Linienstruktur erhalten, die eindeutig das Vorliegen von Struktur- und auch Kristall-wasser zeigt. Allerdings läßt der relativ große Wert des zweiten Moments von 21,0 Gauss^2 vermuten, daß der Kristallwasseranteil größer ist, als es die Strukturformel $K_3[B_3O_3(OH)_6] \cdot H_2O$ angibt. Wahrscheinlicher ist eine Formel wie $\{[K(H_2O)]_3[B_3O_5(OH)_2]\}_n$ für *2*, was einem Schich-ten- und keinem Inselpolyborat entspricht. Der thermische Abbau von *2* in 180 h führt außerdem gerade zu dem in dieser Formulierung um 3 Mole H_2O ärmeren *3*. Wie Abb. 2b zeigt, besteht die erste Ableitung des Protonenresonanzspektrums von *3* auch nur aus einem sehr schmalen Signal mit einer Linienbreite von nur 3,6 Gauss^2. *3* enthält also nur Strukturwasser, was die Formel $\{K_3[B_3O_5(OH)_2]\}_n$ für *3* beweist.

3.4.2. Kaliumtetraborate

Auch die Protonenresonanzspektren der Tetraborate mit Ausnahme von $K_2O \cdot 2\,B_2O_3 \cdot 8\,H_2O$ *(5)* wurden von uns untersucht. *5* hat Menzel [49] durch Lösen von KOH und H_3BO_3 im Molverhältnis 1:1,6 in wenig war-mem Wasser und langsames Abkühlen dieser Lösung dargestellt. *5* scheint das gleiche Polyion zu enthalten wie das $K_2O \cdot 2\,B_2O_3 \cdot 4\,H_2O$ *(6)*.

6 enthält nach der Röntgenstrukturanalyse von Marezio *et al.* [55] das Ion $[B_4O_5(OH)_4]^{2-}$, das schon im Mineral Borax gefunden wurde. Die vollständige Formel von *6* muß demnach $[K(H_2O)]_2[B_4O_5(OH)_4]$ lauten. Atterberg [45] beschrieb bereits eine Darstellungsmethode, eine neuere gibt Ssauka [100] an: eine konz. siedende Lösung von Borsäure wird langsam mit K_2CO_3 neutralisiert und binnen 2 h unter Rühren auf 30 °C abgekühlt. Die Kristalle werden aus Wasser umkristallisiert, mit Alkohol und Äther gewaschen und bei 60—65 °C getrocknet.

Die erste Ableitung der Protonenresonanzkurve von *6* (Abb. 3a) zeigt eine breite, flache Linie ohne besondere Strukturmerkmale. Es folgt daraus, daß *6* sowohl Kristall- als auch Strukturwasser enthalten muß (betr. polykrist. Borsäure vgl. [101] Abb. 5c, S. 236).

Ein weiterer Hinweis auf die Richtigkeit der für *6* vorgeschlagenen Formel $[K(H_2O)]_2[B_4O_5(OH)_4]$ wäre gegeben, wenn das durch 100stün-dige Entwässerung bei 100 °C gebildete $K_2O \cdot 2\,B_2O_3 \cdot 2\,H_2O$ *(7)* nur noch Hydroxidgruppen enthielte. Tatsächlich zeigt das Protonen-resonanzspektrum von *7* (Abb. 3b), daß jetzt nur noch Strukturwasser vorliegt. Die völlige Strukturlosigkeit der schmalen steilen Linie und das zweite Moment von nur 8,5 Gauss^2 lassen keinen anderen Schluß zu. Die Struktur von *7* kann daher mit $K_2[B_4O_5(OH)_4]$ angegeben werden.

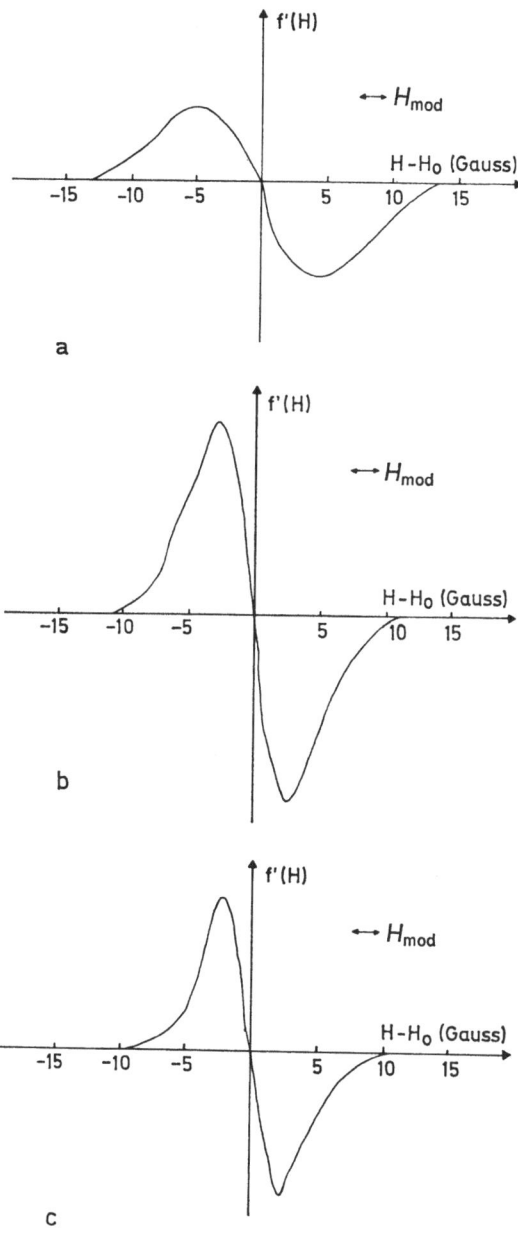

Abb. 3 a—c. Erste Ableitungen der Protonenresonanzabsorptionskurven bei 90 °K

a) $K_2O \cdot 2\,B_2O_3 \cdot 4\,H_2O$; $\Delta\overline{H}^2 = 17,6$ Gauss^2;

b) $K_2O \cdot 2\,B_2O_3 \cdot 2\,H_2O$; $\Delta\overline{H}^2 = 8,5$ Gauss^2;

c) $K_2O \cdot 2\,B_2O_3 \cdot H_2O$; $\Delta\overline{H}^2 = 6,3$ Gauss^2

Das im Vergleich zu *3* ($\Delta \bar{H}^2 = 3,6$ Gauss^2) sehr viel größere zweite Moment von *7* ($\Delta \bar{H}^2 = 8,5$ Gauss^2) läßt sich leicht erklären. Im Inselpolyboration *7* befinden sich vier OH-Gruppen im Molekül und damit in unmittelbarer Nähe zueinander. Jeder magnetische Dipol steht also in Dipol-Dipol-Wechselwirkung mit drei dicht benachbarten Kernen, was zu einer wesentlich größeren Linienverbreiterung führt, als es der Fall ist, wenn nur, wie im Kettenborat von *3*, zwei Kerne in Wechselwirkung stehen.

Daß diese Überlegungen richtig sind, zeigt der weitere thermische Abbau von *7* [102]. Nach weiteren 100 h bei 145 °C ist ein weiteres Mol Wasser abgespalten und die Verbindung $K_2O \cdot 2 B_2O_3 \cdot H_2O$ *(8)* entstanden, der sicher die Formulierung $\{K_2[B_4O_6(OH)_2]\}_n$ zukommt. Nunmehr liegen nur noch zwei OH-Gruppen im Molekül vor. Das zweite Moment von *8* müßte also kleiner sein als das von *7*. Wie Abb. 3c zeigt, ist die Linienbreite tatsächlich geringer, und es wird für $\Delta \bar{H}^2$ ein Wert von 6,3 Gauss^2 gefunden. Beim Erhitzen über 200 °C geht *8* in das wasserfreie $K_2O \cdot 2 B_2O_3$ über, das als $\{K_2B_4O_7\}_n$ zu formulieren ist.

3.4.3. Kaliumpentaborate

In der Gruppe der Kaliumpentaborate sind die zweibasischen Verbindungen nicht so verbreitet wie bei den entsprechenden Natriumpentaboraten. Hier existiert nur $2 K_2O \cdot 5 B_2O_3 \cdot 5 H_2O$ *(9)*, Augers Kaliumborat [66], das bei 115 °C—120 °C aus einer konz. wäßrigen Lösung mit dem Molverhältnis $K_2O:B_2O_3 = 0,4:1$, am besten durch Animpfen, zu erhalten ist. Das Salz scheint das Pentaboration $[B_5O_6(OH)_5]^{2-}$ zu enthalten, das auch in Ezcurrit vorliegt, obwohl es nach seinem Verhalten beim thermischen Abbau besser als $\{[K(H_2O)]_2[B_5O_8(OH)]\}_n$ zu schreiben wäre. Denn beim Erhitzen an der Luft verliert *9* schon unterhalb von 100 °C 4 Mole H_2O, während das letzte erst bei 360—400 °C abgegeben wird, wobei ein amorphes Produkt entsteht [66]. In mit Wasserdampf gesättigter Luft zeigen jedoch thermogravimetrische Untersuchungen und Anwässerungsversuche [13], daß zwischen 40 °C und 140 °C zwei Mole Wasser reversibel ausgetauscht werden. Die zwei weiteren Mole Wasser werden zwischen 140 °C und 360 °C abgebaut, so daß für *9* doch die kristallwasserfreie Formel $K_2[B_5O_6(OH)_5]$ geschrieben werden kann.

Bei den Kalium(1:5)-Boraten ist die Struktur des orthorhombischen, in Wasser schwerlöslichen $K_2O \cdot 5 B_2O_3 \cdot 8 H_2O$ *(10)*, das nur wenig löslicher als $KClO_4$ ist, sowohl durch Röntgenstrukturanalyse [52,103] als auch durch Kernresonanzuntersuchungen [53,54,104] aufgeklärt worden. Die Strukturformel lautet $[K(H_2O)_2][B_5O_6(OH)_4]$. Dargestellt [105] wird *10* durch Abkühlen einer heißen Lösung von KOH und H_3BO_3 im

Molverhältnis 1:3, zweimaliges Umkristallisieren aus viel Wasser, Waschen und Trocknen bei 60 °C. *10* verliert bei 100 °C in 200 h oder bei 125 °C in 80 h in einem Schritt sechs Mole Wasser [102]. Das entstandene Salz hat die Zusammensetzung $K_2O \cdot 5\,B_2O_3 \cdot 2\,H_2O$ *(11)*. Der Versuch, ein Borat der analytischen Zusammensetzung $K_2O \cdot 5\,B_2O_3 \cdot 4\,H_2O$ *(12)* zu erhalten, gelang nur durch Unterbrechung des Erhitzens nach berechnetem Gewichtsverlust.

Die ersten Ableitungen der Protonenresonanzkurve bei 90 °K von *9*, *12* und *11* werden in der Abb. 4 wiedergegeben. Bei *9* liegen deutlich Kristall- und Strukturwasser nebeneinander vor. Das innere Maximum, das für OH-Gruppen charakteristisch ist, ist noch als schwache Schulter zu erkennen. Das zweite Moment von $\Delta\bar{H}^2 = 21$ Gauss² stimmt mit dem von Smith und Richards [53] gemessenen Wert von 22 Gauss² gut überein. Das thermische Abbauprodukt *12* enthält, wie Abb. 4b zeigt, trotz des Abbaus von 2 Mol Wasser aus *9* immer noch Kristallwasser (Schulter der Kurve). *12* ist keine stabile Hydratstufe des Pentaborats. Offenbar folgt der Abspaltung des Kristallwassers fast gleichzeitig eine Kondensation zwischen Hydroxidgruppen verschiedener Moleküle.

Das nach weiterem thermischen Abbau erhältliche *11* besitzt, wie die Linienstruktur in Abb. 4c zeigt, nur noch Strukturwasser. Das zweite Moment von $\bar{H}^2 = 7{,}0$ Gauss² spricht für die Formulierung als Kettenpolyborat $\{K[B_5O_7(OH)_2]\}_n$.

Erst bei 450 °C ist die Dehydratisierung nahezu vollständig. Das wasserfreie $K_2O \cdot 5\,B_2O_3$ enthält nach Kristallstrukturuntersuchungen von Krogh-Moe [106,107], der eine α-Hochtemperaturform mit kleiner Dichte und eine β-Tieftemperaturform mit größerer Dichte unterscheidet, senkrecht zueinander stehende Pentaboratringe, die über die außenstehenden Sauerstoffatome zu Helixketten verbunden sind; die Helixketten bilden mit Nachbarketten Schichten; die Strukturformel ist am besten mit $\{KB_5O_8\}_n$ zu beschreiben.

3.4.4. Durch Boresterhydrolyse dargestellte Kaliumpolyborate

Durch Kristallisation [108] nach vorsichtiger Hydrolyse von Borsäuretrimethylester in Gegenwart von Kaliumalkoxiden wurden je nach den Konzentrationsverhältnissen der Reaktionspartner aus organischen Lösungsmitteln folgende Kaliumpolyborate erhalten:

$$K_2O \cdot 12\,B_2O_3 \cdot 7\,H_2O \quad (13);$$
$$K_2O \cdot 5\,B_2O_3 \cdot 4\,H_2O \quad (12);$$
$$K_2O \cdot 3\,B_2O_3 \cdot 4\,H_2O \quad (14),$$
$$2\,K_2O \cdot 5\,B_2O_3 \cdot 5\,H_2O \quad (9) \text{ und}$$
$$K_2O \cdot 2\,B_2O_3 \cdot 2\,H_2O \quad (7).$$

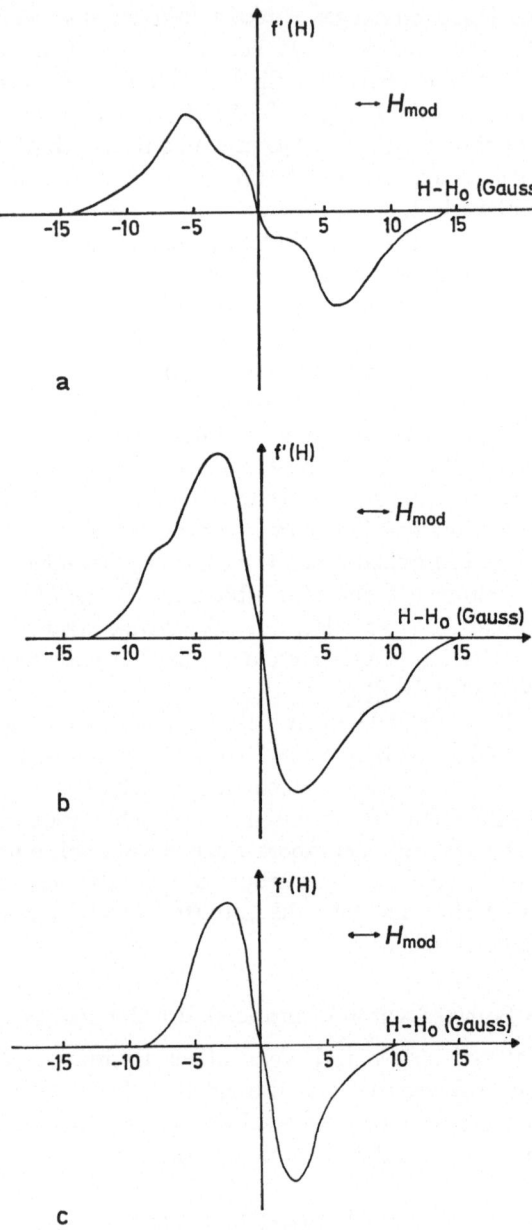

Abb. 4 a—c. Erste Ableitungen der Protonenresonanzabsorptionskurven bei 90 °K

a) $K_2O \cdot 5 B_2O_3 \cdot 8 H_2O$ mit $\Delta \overline{H}^2 = 21,0$ Gauss²;
b) $K_2O \cdot 5 B_2O_3 \cdot 4 H_2O$ mit $\Delta \overline{H}^2 = 15,0$ Gauss²;
c) $K_2O \cdot 5 B_2O_3 \cdot 2 H_2O$ mit $\Delta \overline{H}^2 = 7,0$ Gauss²

Dabei wurde die Verbindung *13* unter extremen Bedingungen gefällt (s. unten). Tabelle 3 zeigt die optimalen Konzentrationsverhältnisse der Reaktionspartner bei der Bildung der Polyborate *12, 14, 9* und *7*. Ihr Kondensationsgrad hängt, wie bei den Ammoniumpolyboraten beschrieben, vom Wassergehalt, vom Kalium- zu Bor-Verhältnis und vom Lösungsmittel ab.

Tabelle 3. *Konzentrationsverhältnisse der bei der üblichen Boresterhydrolyse gefällten Kaliumpolyborate*

		Konz.-Verhältnisse			Lsgsmittel
		im Salz K:B	im Reaktionsgemisch K:B	H_2O	
$K_2[B_4O_5(OH)_4]$	*(7)*	1:2	1:0,9	1%	Äthanol
$K_2[B_5O_6(OH)_5]$	*(9)*	1:2,5	1:3,2	2%	Äthanol
$K[B_3O_3(OH)_4]$	*(14)*	1:3	1:11,2	4%	Testbenzin/ Äthanol
$K[B_5O_6(OH)_4]$	*(12)*	1:5	1:11,2	2,5%	Testbenzin/ Äthanol

Bei den Kaliumsalzen ist es prinzipiell möglich, die Erfahrungstatsache, daß die durch Esterhydrolyse erhaltenen Salze kristallwasserfrei sind, zu überprüfen. Die Kaliumpolyborate enthalten im Kation keine Wasserstoffatome und könnten daher Protonenresonanzuntersuchungen unterworfen werden. Da aber die durch Esterhydrolyse erhaltenen Kaliumpolyborate trotz Trocknen im Vakuum bei erhöhter Temperatur immer noch wasserstoffhaltiges Lösungsmittel enthielten und deshalb für die Protonenresonanzuntersuchungen ungeeignet blieben, wurde versucht, das Triborat aus wasserstoff-freiem CCl_4 zu fällen. Zu diesem Zweck wurde versucht, Kaliummethylat in CCl_4 aufzulösen. Beim Schütteln gab es jedoch eine heftige Reaktion, bei der Kaliumchlorid ausfiel. Nach dem Abfiltrieren waren in der CCl_4-Lösung jedoch noch Kalium-Ionen vorhanden. Die Lösung wurde mit der konz. Boresterlösung in CCl_4 gemischt, und eine definierte Menge Wasser wurde langsam zugetropft. Nach längerem Stehen fiel ein Salz mit dem für Borate ungewöhnlichen Verhältnis von Kalium:Bor $= 1:12$ aus. Die Analysenwerte entsprechen der Zusammensetzung $K_2O \cdot 12\,B_2O_3 \cdot 7\,H_2O$ und der Anwesenheit von ca. 2% organischem Lösungsmittel. Da sich die genauen Anteile an Lösungsmittel analytisch nicht ermitteln lassen, kann es sich auch um $K_2O \cdot 12\,B_2O_3 \cdot 5\,H_2O$ handeln; dann müßten aber noch nach dem Trocknen knapp 5% organisches Lösungsmittel vorhanden sein. Ein Wert von $6\,H_2O$ in der Verbindung käme ihren C-Analysen am nächsten, wenn man annimmt, daß das adsorbierte Lösungsmittel nur CCl_4 und auch nicht durch Verseifung des Borsäuretrimethylesters entstandenes Methanol ist. Es müßten dann aber in dieser Verbindung 24fach aggregierte Anionen vorkommen, was kaum anzunehmen ist; ein Dodekaborat, also die Mindestformulierung, wäre wahrscheinlicher.

Die Darstellung des für Protonenresonanzuntersuchungen gewünschten Triborats *(14)* in protonenfreiem CCl_4 gelang, als statt Kaliummethylat eine Lösung von Kalium-n-butylat in Tetrachlorkohlenstoff als Ausgangskomponente benutzt wurde.

3.4.5. Eigenschaften der Kaliumpolyborate

Alle aus organischem Lösungsmittel gewonnenen Kaliumpolyborate sind außerordentlich hygroskopisch. Sie lösen sich in Wasser, wobei sie in K^+, $B(OH)_4^-$ und $B(OH)_3$ zerfallen. In organischen Lösungsmitteln sind sie unlöslich; versucht man, sie in Dimethylsulfoxid zu lösen, so quellen sie, bleiben aber ungelöst. Die Kriställchen laden sich leicht elektrostatisch auf und bleiben an Glaswänden haften.

Nach Röntgen-Diffraktionsaufnahmen sind alle diese Polyborate kristallin. Für das bisher unbekannte Kaliumpolyborat *(14)* betragen die wichtigsten Werte für die röntgenographischen d-Abstände in Å: 12,55 (100); 7,20 (80); 6,22 (100); 4,70 (30); 3,59 (40); 3,26 (30) und 3,00 (40). In Klammern stehen die geschätzten Intensitäten I/I_1 in %, bezogen auf die beiden stärksten Banden mit 100%.

Auch nach dem Trocknen im Wasserstrahlpumpenvakuum bei 80 °C bis 110 °C enthalten alle neuen Kaliumpolyborate noch etwas Lösungsmittel.

Bei den Verbindungen *9* und *7* sind sogar etwa 3% einer kohlenstoffhaltigen Substanz, wahrscheinlich des Lösungsmittels, analytisch zu erfassen; bei den anderen Boraten sind die für das Lösungsmittel charakteristischen C-Werte geringer. Beim Trocknen im Vakuum bei erhöhter Temperatur werden die Kaliumpolyborate mehr oder weniger röntgenamorph; das Borat *14* behält aber zum Beispiel seine Bande bei 12,55 Å, wenn auch nicht mehr in der ursprünglichen Intensität. Die nun röntgenamorphen Verbindungen besitzen in ihren Remissionsspektren zwischen $\lambda = 600$ und 2600 nm nicht wie die als Vergleichssubstanzen eingesetzten Verbindungen $Na_2B_4O_7 \cdot 10\,H_2O$ und $K_2B_4O_7 \cdot 4\,H_2O$ ausgeprägte H_2O-Frequenzen.

Beim Abbau des im Exsikkator bei Raumtemperatur über Silikagel und Paraffinschnitzeln getrockneten Kaliumpolyborates *7* wurde mit Hilfe der Differentialthermoanalyse beobachtet, daß bis 500 °C ohne erkennbare Stufen 20% der ursprünglichen Substanzmenge abgebaut werden. Der bei 500 °C erhaltene schwärzliche Rückstand enthält 29,5% K und 16,4% B, ist also vermutlich das durch Kohlenstoff verunreinigte Kettenpolyborat $\{K_2[B_4O_6(OH)_2]\}_n$, das auch durch thermischen Abbau bei 145 °C aus dem aus wäßriger Lösung gewonnenen $K_2O \cdot 2\,B_2O_3 \cdot 4\,H_2O$ *(6)* erhältlich ist. Der weitere Abbau scheint zu einem Gemisch von Kaliummetaborat und Metaborsäure zu führen.

3.4.6. Strukturen der Kaliumpolyborate

Alle fünf durch Boresterhydrolyse dargestellten Kaliumpolyborate besitzen gegenüber allen aus wäßrigen Lösungen zu gewinnenden Polyboraten Besonderheiten. Während *13* und *14* bisher im System der Kaliumpolyborate vollständig unbekannt waren und *12* bisher nicht als reine Verbindung existiert, sondern nur durch berechnete Unterbrechung des thermischen Abbaus von *10* gewonnen wird und seinem Protonenresonanzspektrum nach keine einheitliche Phase zu sein scheint, entsprechen *9* und *7* auf den ersten Blick und den Summenformeln nach den aus wäßriger Lösung gewonnenen Verbindungen. Während *9* und *7* aus dem organischen Lösungsmittel bei Raumtemperatur auskristallisieren, entstehen sie aus wäßriger Lösung in der Siedehitze (bei 114 °C—115 °C). Dennoch spricht nichts dagegen, beide kristallwasserfrei zu formulieren, nämlich *9* als $K_2[B_5O_6(OH)_5]$ und *7* als $K_2[B_4O_5(OH)_4]$. Die der Verbindung *12* analoge Ammoniumverbindung $NH_4[B_5O_6(OH)_4]$ wurde ebenfalls durch Esterhydrolyse [41] dargestellt. *12* enthält im Gegensatz zu dem aus wäßriger Lösung gewonnenen *10*, das durch Strukturuntersuchungen [52,53,54,103,104] genau bekannt und als $K(H_2O)_2[B_5O_6(OH)_4]$ zu formulieren ist, kein Kristallwasser und ist daher ein Inselpolyborat mit der Strukturformel $K[B_5O_6(OH)_4]$.

Da Wegener [109] für die Verbindungen *13*, $K_2O \cdot 12\,B_2O_3 \cdot 7\,H_2O$, und *14*, $K_2O \cdot 3\,B_2O_3 \cdot 4\,H_2O$, die aus dem protonenfreien Lösungsmittel CCl_4 gewonnen wurden, durch Protonenresonanzuntersuchungen gezeigt hat, daß sie tatsächlich kristallwasserfrei sind, kann man für die analog *14* dargestellten Verbindungen *9*, *7* und *12* das gleiche annehmen.

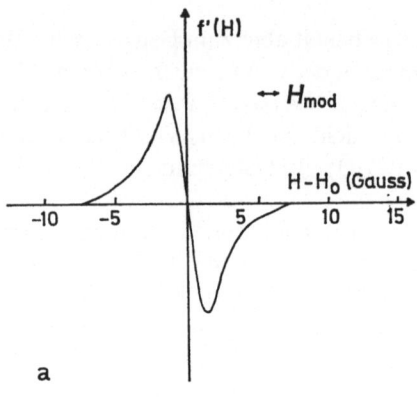

a

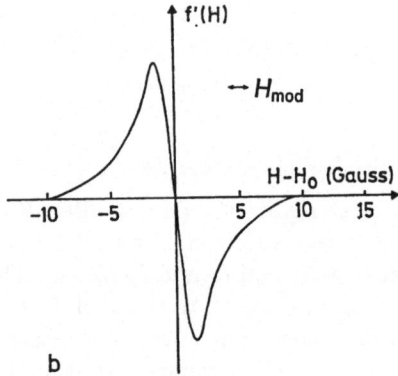

b

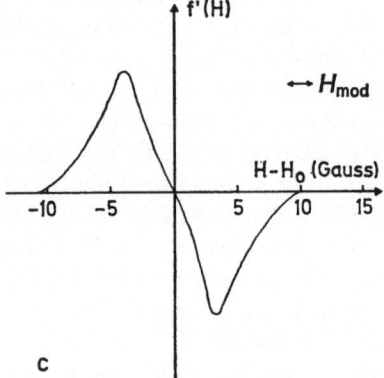

c

Abb. 5 a—c. Erste Ableitungen der Protonenresonanzabsorptionskurven bei 90 °K

a) Kaliumpolyborat *(14)* mit $\Delta\bar{H}^2 = 4{,}1$ Gauss^2;

b) Kaliumpolyborat *(13)* mit $\Delta\bar{H}^2 = 6{,}7$ Gauss^2;

c) Borsäure (298 °K) mit $\Delta\bar{H}^2 = 11$ Gauss^2 [101]

Wie Abb. 5 zeigt, bestätigen die Linienstrukturen der ersten Ableitungen der bei 90 °K aufgenommenen PMR-Absorptionskurven der Verbindungen *13* und *14*, daß diese Polyborate Wasser ausschließlich in Form von Hydroxidgruppen und nicht in Form von Kristallwasser enthalten. Auch die zweiten Momente von *13* mit $\Delta \bar{H}^2 = 6,7$ Gauss^2 und von *14* mit $\Delta \bar{H}^2 = 4,1$ Gauss^2 unterstreichen diese Feststellung [109]. Da sich die Linienbreiten der Protonenresonanzabsorptionslinien des Polyborats *13* mit $\delta H = 2,5$ Gauss und der Borsäure mit $\delta H = 9$ Gauss (Abb. 5b und c) stark unterscheiden, ist es unwahrscheinlich, daß es sich bei der durch Boresterhydrolyse gewonnenen Verbindung *13* um Borsäure handelt, an die K_2O in geringen Mengen adsorbiert ist. Es kann sich bei Verbindung *13* nur um $K[B_{12}O_{15}(OH)_7]$, vielleicht auch um $K[B_{12}O_{16}(OH)_5]$ handeln.

Für das Kaliumpolyborat *14* muß eindeutig die Strukturformel $K[B_3O_3(OH)_4]$ lauten.

3.4.7. Kaliummagnesiumborate

Das Mineral *Kaliborit*, für das früher die Formel $K_2O \cdot 4 \, MgO \cdot 11 \, B_2O_3 \cdot 18 \, H_2O$ angegeben wurde, sollte nach Voraussagen von Christ [6] eine Doppelverbindung der Strukturformel $K[B_5O_6(OH)_4] \cdot 2 \, Mg[B_3O_3(OH)_5]$ $\cdot 2 \, H_2O$ sein. Neuere Untersuchungen von Corazza und Sabelli [110] ergaben aber für das Mineral die Zusammensetzung

$$KMg_2HB_{12}O_{16}(OH)_{10} = K_2O \cdot 4 \, MgO \cdot 12 \, B_2O_3 \cdot 19 \, H_2O \, .$$

Die Struktur wurde mit Hilfe von dreidimensionaler Patterson-Projektion und dreidimensionaler Fouriersynthese bestimmt und verfeinert. Danach setzt sie sich aus Kettenpolyborationen $\{B_6O_8(OH)_5\}_n^{3n-}$ zusammen, die Doppelringe bilden und entlang der kristallographischen c-Achse durch sich abwechselnde BO_4-Tetraeder und BO_3-Dreiecke verknüpft sind. Die Mg-Ionen haben oktaedrische, die K-Ionen kubisch verzerrte Koordination, wobei die Kationen die Brücken zwischen den B—O-Ketten bilden.

3.5. Rubidium- und Caesiumborate

Rubidium: Das System $Rb_2O/B_2O_3/H_2O$ wurde erstmals von Rollet und Andres [111] bei 18 °C und später von Tolédano [94] bei 80 °C untersucht.

Die aus den Lösungen bei verschiedenen Temperaturen isolierten Borate haben folgende Zusammensetzungen:

$$Rb_2O \cdot B_2O_3 \cdot 8 H_2O \quad (1),$$

$$Rb_2O \cdot B_2O_3 \cdot 2,5 H_2O \text{ oder } Rb_2O \cdot B_2O_3 \cdot 2,67 H_2O \quad (2),$$

$$Rb_2O \cdot B_2O_3 \cdot 0,5 H_2O \text{ oder } Rb_2O \cdot B_2O_3 \cdot 0,67 H_2O \quad (3),$$

$$Rb_2O \cdot 2 B_2O_3 \cdot 3 H_2O \quad (4),$$

$$2 Rb_2O \cdot 5 B_2O_3 \cdot 6 H_2O \quad (5) \text{ und}$$

$$Rb_2O \cdot 5 B_2O_3 \cdot 8 H_2O \quad (6).$$

1 existiert nur bei relativ tiefen Temperaturen und geht bei etwa 20 °C in *2* über. Über Zusammensetzung und Struktur von *2* und *3* gibt es die gleichen Diskussionen wie bei den entspr. Kaliummetaboraten. Früher [112)] wurde *2* auch als $Rb_2O \cdot B_2O_3 \cdot 3 H_2O$ formuliert. Heute halten Lehmann u. Mitarb. [113,114)] *2* für $Rb_2O \cdot B_2O_3 \cdot 2,67 H_2O$ mit der Strukturformel $[Rb_3(H_2O)][B_3O_3(OH)_6]$ und Tolédano u. Mitarb. [94)] *2* für $Rb_2O \cdot B_2O_3 \cdot 2,5 H_2O$. Beim thermischen Abbau [94)] geht *2* bei 190 °C in $Rb_2O \cdot B_2O_3 \cdot 0,5 H_2O$ bzw. $Rb_2O \cdot B_2O_3 \cdot 0,67 H_2O$ [113)] über. *4* ist bei Raumtemperatur eine stabile Phase und ergibt beim Eindampfen hygroskopische prismatische Kristalle [112)]. *4* zersetzt sich bei 172 °C zum 2-Hydrat, bei 244 °C zum 1-Hydrat und bei 320 °C zu $Rb_2O \cdot 2 B_2O_3$.

5 wurde als neue stabile Phase in der Isotherme bei 80 °C gefunden. Es geht zwischen 200 °C und 214 °C in das 3-Hydrat, zwischen 240 °C und 255 °C in das 2-Hydrat und bei 450 °C in $2 Rb_2O \cdot 5 B_2O_3$ über, das schwach röntgenkristallin ist.

Das orthorhombisch kristallisierende *6* ist schon lange [105)] bekannt. Bei seinem thermischen Abbau erhält man bei 166 °C das 2-Hydrat, bei 262 °C das 1-Hydrat und bei 327 °C $Rb_2O \cdot 5 B_2O_3$. Analog zu den Kaliumpolyboraten fällt die Aufstellung der Strukturformeln von *2* und *3* schwer, während *1* als $Rb(H_2O)_2[B(OH)_4]$, *4* als $[Rb_2(H_2O)]$ $[B_4O_5(OH)_4]$ und *6* als $[Rb(H_2O)_2][B_5O_6(OH)_4]$ formuliert werden können. Eine Strukturformel $[Rb_2(H_2O)_{0,5}][B_5O_6(OH)_5]$ für *5* ist sehr unwahrscheinlich.

Cäsium: Auch vom schon bei 18 °C untersuchten System $Cs_2O/B_2O_3/$ H_2O [111)] wurden kürzlich von Kocher [115)] alle Isothermen zwischen −5 °C und 100 °C aufgenommen. Danach existieren, ähnlich wie bei den Rubidiumboraten, folgende Verbindungen:

$$Cs_2O \cdot B_2O_3 \cdot 8 H_2O \quad (1, \alpha\text{- und } \beta\text{-Form}),$$

$$Cs_2O \cdot B_2O_3 \cdot 2 H_2O \quad (2),$$

$$Cs_2O \cdot 2 B_2O_3 \cdot 5 H_2O \quad (3),$$

$$Cs_2O \cdot 2 B_2O_3 \cdot 2 H_2O \quad (4),$$

$$2 Cs_2O \cdot 5 B_2O_3 \cdot 7 H_2O \quad (5) \text{ und}$$

$$Cs_2O \cdot 5 B_2O_3 \cdot 8 H_2O \quad (6).$$

Der entscheidende Unterschied zwischen Rubidium- und Caesium-boraten liegt bei den Metaboraten, denn die analytische Zusammensetzung von *2* ist eindeutig, und seine Struktur wird mit $\{Cs[BO(OH)_2]\}_n$ formuliert [113]. *1* und *2* wurden bereits von Lehmann und Gaube [114] gefunden. Die α-Form von *1* ist bis 50 °C stabil und geht dann in die β-Form über, die bei 83 °C *2* gibt, das seinerseits bei 170 °C gleich zu $Cs_2O \cdot B_2O_3$ wird. Für die α-Form von *1* wird die Formel $Cs(H_2O)_2$ $[B(OH)_4]$ und für die β-Form die kondensierte Struktur $\{Cs(H_2O)_3$ $[BO(OH)_2]\}_n$ vorgeschlagen [113]; damit ist der weitere Abbau zu *2* gut zu erklären.

3 wird thermisch bei 110 °C zu *4* abgebaut. Das ab 45 °C stabile *5* geht zwischen 110 °C und 150 °C in das 3-Hydrat über, das zwischen 260 °C und 400 °C wasserfrei wird. *6* gibt zwischen 70 °C und 115 °C sechs Mole Wasser ab; das entstandene $Cs_2O \cdot 5 B_2O_3 \cdot 2 H_2O$ ist bis 400 °C beständig. Die Strukturen entsprechen denen der Rubidiumpolyborate.

Schon Reischle [116] hat ein Caesiumborat durch Zugabe von alkoholischer Borsäurelösung zu einer Lösung von Cs_2O in absolutem Äthanol und Eindunsten in Form eines feinen kristallinen Niederschlages erhalten; dieses Caesiumborat soll in absolutem Äthanol ziemlich leicht löslich sein. Zur Analyse bestimmte Reischle damals nur den Cs-Wert zu 57,11% $Cs_2O = 53,87\%$ Cs und schloß aus diesem Wert allein auf die Formel $Cs_2O \cdot 3 B_2O_3$.

Man kann sich der Annahme nicht entziehen, daß er damals schon $Cs_2O \cdot B_2O_3 \cdot 2 H_2O$ dargestellt hat. Bei unseren Versuchen, Borsäuretrimethylester in Gegenwart einer 0,1 n Lösung von Cs_2CO_3 in absolutem Methanol zu hydrolysieren, wurden nämlich nach Einengen und Abfiltrieren der zuerst ausfallenden, schwerer löslichen Borsäure auch bei Anwendung verschiedener Konzentrationsverhältnisse immer Caesiumborate mit dem Cs:B-Verhältnis 1:1 erhalten. Diese enthielten bei Raumtemperatur 48—49% Cs und 4,2—4,4% B; nach dem Trocknen im Vakuum bei 50 °C enthielt eines dieser Salze 60,54% Cs; 5,11% B; 2,00% H und 6,06% C. Dieses Ergebnis entspricht den zu erwartenden Werten für ein etwas Lösungsmittel enthaltendes Caesiummonoborat $Cs_2O \cdot B_2O_3 \cdot 2 H_2O = Cs[B(OH)_4]$, dessen theoretische Werte 62,76% Cs; 5,11% B und 2,01% H lauten. Da die Verbindung stark feuchtigkeits- und kohlendioxidempfindlich ist, ist es denkbar, daß Reischle schon 1893 diese Verbindung in der Hand hatte.

Beim Versuch, das vorstehend beschriebene, stark alkalische Reaktionsgemisch durch Reaktion in Eisessig statt in Methanol oder durch Fällung mit wenig 7%iger Salzsäure statt durch Fällen mit wenig Wasser saurer zu machen, fielen Gemische von $Cs[B(OH)_4]$, Borsäure und Caesiumacetat bzw. CsCl aus.

Die durch Boresterhydrolyse erhaltene Substanz, für die die Strukturformel $Cs[B(OH)_4]$ angenommen werden muß, zersetzt sich beim Aufnehmen eines Röntgenpulverdiagramms. Sie hat im IR-Spektrum nur wenige ausgeprägte Banden, nämlich bei 832, 942, 984 und 1063 bis 1073 cm^{-1}, bei 1194, 1317, 1442 und 1665 cm^{-1} sowie bei 2823, 2930 und 3423 cm^{-1}. Während erstere klar für das Vorliegen von vierbindigem Bor sprechen, könnte man die Banden um 1070, bei 1317 und 1442 cm^{-1} auch dreibindigem Bor zuordnen; bei 1665 cm^{-1} ist die δ (OH)-, bei 3423 cm^{-1} die ν (OH)-Schwingung zu erwarten.

In den wasserfreien Systemen Rb_2O/B_2O_3 und Cs_2O/B_2O_3 tritt [115,117] die gleiche Polyboratreihe auf wie bei K_2O/B_2O_3, nämlich $M_2O \cdot B_2O_3$ (zwei Kristallformen mit Rb), $M_2O \cdot 2\,B_2O_3$, $2\,M_2O \cdot 5\,B_2O_3$, $M_2O \cdot 3\,B_2O_3$, $M_2O \cdot 4\,B_2O_3$, $M_2O \cdot 5\,B_2O_3$ (je zwei Formen mit Rb und Cs) sowie $Cs_2O \cdot 9\,B_2O_3$ (zwei Formen). Von Krogh-Moe wurden die Strukturen von $Cs_2O \cdot 3\,B_2O_3$ [58] und von α-$Cs_2O \cdot 9\,B_2O_3$ [118] aufgeklärt. In Ersterem bildet das polymere Anion $\{B_3O_5\}_n^{n-}$ ein dreidimensionales Gerüst aus sechsatomigen Ringen mit alternierenden B- und O-Atomen, die über Sauerstoffbrücken zu Helices werden. In letzterem sind Sechsringe mit dreibindigem Bor über Sauerstoffbrücken miteinander und mit den Cs$^+$-Kationen verbunden.

3.6. Beryllium- und Magnesiumborate

3.6.1. Berylliumborate

Ein bekanntes wasserhaltiges Berylliumborat ist das Mineral Hambergit, $4\,BeO \cdot B_2O_3 \cdot H_2O$; es gibt seinen Wassergehalt nur bei hoher Temperatur ab. Eine Verbindung dieser Formel läßt sich auch synthetisch darstellen [119], ebenso ein Mono- und ein Trihydrat [120]. Die Struktur von Hambergit wurde durch Kristallstrukturanalyse aufgeklärt [121]; im Molekül sind die Be-Ionen tetraedrisch von drei O-Atomen und einer OH-Gruppe, die B-Atome von drei O-Atomen in Form eines gleichseitigen Dreiecks umgeben. Die Tetraeder und Dreiecke berühren sich nur in den Ecken. Ein O-Atom gehört zu 2 Be und 1 B, eine OH-Gruppe nur zu 2 Be, so daß als Strukturformel $Be_2OH(BO_3)$ geschrieben werden kann.

Wasserfrei gibt es nur $3\,BeO \cdot B_2O_3$, das nadelförmig im monoklinen Kristallsystem kristallisiert.

3.6.2. Wasserhaltige Magnesiumborate

Folgende reine Magnesiumboratmineralien sind bekannt:

$9\,MgO \cdot B_2O_3 \cdot 8\,H_2O$, Wightmanit;

$2\,MgO \cdot B_2O_3 \cdot H_2O$, Ascharit oder Szaibelyit;

$MgO \cdot B_2O_3 \cdot H_2O$, Pinnoit;

$2 MgO \cdot 3 B_2O_3 \cdot 15 H_2O$, Kurnakovit und Inderit bzw. Lesserit;

$3 MgO \cdot 5 B_2O_3 \cdot 4^1/_3 H_2O$ oder $4 MgO \cdot 7 B_2O_3 \cdot 4 H_2O$, Preobrazhenskit;

$2 MgO \cdot 4 B_2O_3 \cdot 5 H_2O$, Halurgit;

$MgO \cdot 3 B_2O_3 \cdot 5 H_2O$, Aksait und

$MgO \cdot 4 B_2O_3 \cdot 4 H_2O$, Paternoit.

Im System $MgO/B_2O_3/H_2O$ existieren außerdem die synthetisch zu gewinnenden Polyborate $2 MgO \cdot B_2O_3 \cdot 3 H_2O$, $MgO \cdot 2 B_2O_3 \cdot 8^1/_2 H_2O$ oder $MgO \cdot 2 B_2O_3 \cdot 9 H_2O$ sowie $MgO \cdot 3 B_2O_3 \cdot 5 H_2O$, $MgO \cdot 3 B_2O_3 \cdot 6 H_2O$, $MgO \cdot 3 B_2O_3 \cdot 7 H_2O$ und $MgO \cdot 3 B_2O_3 \cdot 7^1/_2 H_2O$.

Während von *Wightmanit* außer der Zusammensetzung nichts bekannt ist [122], löst sich *Ascharit* erst beim Sieden in 0,1 n Mineralsäure und gibt sein Wasser erst bei 615—660 °C ab [123]. Es ist als Monohydrat nur unter Druck oder im geschlossenen Rohr bei 150—190 °C, also unter hydrothermalen Bedingungen, zu synthetisieren. Nach seinem [11]B—NMR-Spektrum enthält es nur vierfach koordinierte Boratome [124].

Kristallstrukturuntersuchungen [125,126] an *Pinnoit* zeigen, daß die Boratome tetraedrisch von Sauerstoff umgeben sind und daß es ein Inselpolyborat mit der Strukturformel $Mg[B_2O(OH)_6]$ ist. Es ist auch synthetisch im System $MgO/B_2O_3/H_2O$ bei 100 °C zu isolieren.

Das triklin kristallisierende *Kurnokavit* wurde früher mit 13 Mol H_2O formuliert; es unterscheidet sich vom monoklinen *Inderit* nur durch seine Kristallklasse. In der Literatur wurde Kurnakovit oft Inderit genannt und Inderit dann *Lesserit*. Die Kristallstrukturanalysen von Inderit unter dem Namen Lesserit [127] und von Kurnakovit [128] zeigen, daß in beiden das Inselpolyboration $[B_3O_3(OH)_5]^{2-}$ vorliegt, was auch [11]B—NMR-Untersuchungen bestätigen [129,130]. Zur besseren Unterscheidung wird Inderit als $\alpha\text{-}Mg[B_3O_3(OH)_5] \cdot 5 H_2O$ und Kurnakovit als $\beta\text{-}Mg[B_3O_3(OH)_5] \cdot 5 H_2O$ bezeichnet.

Die Strukturen der übrigen Mineralien sind noch unbekannt.

Preobrazhenskit gibt bei 515—550 °C alles Wasser auf einmal ab [123] und besitzt nach seinen ^{11}B-NMR-Spektren sowohl drei- als auch vierbindiges Bor [124]. Von *Halurgit* [131] sind bisher nur die Formel und die Gitterdaten bekannt. Daran, daß Halurgit mit $MgO \cdot 3 B_2O_3 \cdot 5 H_2O$ [132] identisch ist, wird nach Vergleich der IR-Spektren gezweifelt [133]. Für *Paternoit* wird das Vorliegen des Anions $[B_4O_4(OH)_5]^-$ vermutet.

$2 MgO \cdot B_2O_3 \cdot 3 H_2O$ ist nur ein wasserreicheres Homologes von Ascharit; im Gleichgewichtssystem erscheint es bei der Isotherme von 83 °C nur in Gegenwart von überschüssigem $MgCl_2$. $MgO \cdot 2 B_2O_3 \cdot 9 H_2O$ [134] ist wasserreicher als Halurgit und schwer zu isolieren, da es in Lösung gegenüber Inderit instabil ist. Es kristallisiert auch wie Inderit monoklin. Ihm wird die Strukturformel $Mg[B_4O_4(OH)_6] \cdot 6 H_2O$ zugeschrieben [135] mit einem Anion, das sich aus dem Anion des Inderit,

$[B_3O_3(OH)_5]^{2-}$, rein formal mit der Ersetzung eines H-Atoms durch die BO(OH)-Gruppe aufbauen läßt. Diese Formulierung erklärt seinen leichten Übergang in Inderit.

Im wäßrigen System kristallisiert $MgO \cdot 3 B_2O_3$ je nach Temperatur mit $7^1/_2$, 7, 6 oder 5 Molen Wasser. Während die Verbindung mit 7,5 Mol H_2O schon bei Raumtemperatur im System $MgO/B_2O_3/H_2O$ als Gleichgewichtskomponente auftritt, sind die übrigen Verbindungen erst bei höheren Temperaturen beständig. Von diesen drei Stufen ist nur die mit 5 Mol H_2O stabil; die beiden anderen wandeln sich unter der Mutterlauge langsam in das 5-Hydrat um [113].

Anhand der IR-Spektren dieser Hydrate von $MgO \cdot 3 B_2O_3$ diskutieren Lehmann und Kessler [133] den Begriff „Hydrat" in der Borchemie und unterscheiden bei bekannter Konstitution für anionengleiche Borate zwischen (reversiblen) Hydratstufen und (irreversiblen) Konstitutionshydraten. Diese Unterscheidung hat vorläufigen Charakter, insbes. wenn genaue Strukturformeln nicht bekannt sind.

3.6.3. Wasserfreie Magnesiumborate

Im wasserfreien System existieren Verbindungen mit den $MgO:B_2O_3$-Verhältnissen 1:1, 2:1 und 3:1. Die 1:1-Verbindung $\{Mg(BO_2)_2\}_n$ entsteht [136] bei 580 °C exotherm aus den Komponenten. Sie gibt bei längerem Erhitzen auf über 800 °C B_2O_3 ab. Das Anion der Verbindung ist ringförmig wie $B_3O_6^{3-}$ in $Na_3B_3O_6$ oder kettenförmig wie BO_2^{2-} in $\{Ca(BO_2)_2\}_n$ gebaut. Die 2:1-Verbindung existiert sowohl monoklin als auch triklin; für die monokline Form ist durch Röntgenpulveraufnahmen die Anwesenheit von $B_2O_5^{4-}$-Gruppen bewiesen [137]. Die 3:1-Verbindung ist isomorph mit der entsprechenden Co-Verbindung und enthält nur dreibindiges Bor mit dem Anion BO_3^{3-} [138].

Das Mineral *Boracit* ist wasserfreies $5 MgO \cdot 7 B_2O_3 \cdot MgCl_2$. Es enthält nach Kristallstrukturuntersuchungen [139] B_5- und B-Gruppen in schichtenförmiger Vernetzung.

3.6.4. Magnesium-Calcium(Strontium)-Borate

Von den natürlich vorkommenden Mineralien, die Mg- und Ca-Polyborate enthalten, sind *Inderborit*, $CaO \cdot MgO \cdot 3 B_2O_3 \cdot 11 H_2O$, und *Hydroborazit*, $CaO \cdot MgO \cdot 3 B_2O_3 \cdot 6 H_2O$, am bekanntesten. Beide kristallisieren im monoklinen System, und von beiden sind die Kristallstrukturen bekannt. Während Inderborit [140] das Inselpolyboration $[B_3O_3(OH)_5]^{2-}$ enthält, ist in Hydroborazit [141] das Kettenpolyboration $\{B_3O_4(OH)_3\}_n^{2n-}$ die immer wiederkehrende Struktureinheit.

Wie bei den Paaren Borax/Kernit, Ulexit/Probertit und Inyoit/Colemanit, ist auch beim Paar Inderborit/Hydroborazit das wasserärmere

Mineral ein Kettenpolyborat. Bei seiner Entstehung muß infolge von erhöhtem Druck oder erhöhter Temperatur ein Wassermangel vorgelegen haben, sonst wäre jeweils das unter Normalbedingungen beständigere Inselpolyborat entstanden.

Strontioborit ist das einzige Magnesium-Strontiumborat natürlichen Vorkommens, in dem Sr isomorph durch Ca bis zum Verhältnis 3:1 ersetzt sein kann. Nach Kondratjewa [142], die Einkristallröntgenaufnahmen des Minerals gemacht und die die Kristallparameter bestimmt hat, ist die von Lobanowa [143] ermittelte chemische Formel 4 (Sr, Ca)O · 2 MgO · 12 B_2O_3 · 9 H_2O nicht mit den Röntgendaten (Dichte, Raumgruppe) vereinbar.

3.7. Calcium-, Strontium- und Bariumborate

3.7.1. Wasserhaltige Calciummonoborate

Das erdalkalireiche 2 CaO · B_2O_3 · H_2O (*1*; [144]) bildet sich unter anderem beim Erhitzen wäßriger Lösungen von $Ca(OH)_2$ und H_3BO_3 im Molverhältnis 2:1 bei p_H 12 auf 180 °C. Es wird als sekundäres Calciummonoborat, $\{Ca[BO_2(OH)]\}_n$, formuliert. Es gibt sein Wasser zwischen 200 ° und 400 °C ab unter Bildung von $Ca_2B_2O_5$. Röntgenpulveraufnahmen von *1* und $Ca_2B_2O_5$ sind von den gleichen Autoren veröffentlicht, außerdem kürzlich [145] die Röntgendaten. *1* kristallisiert auch unter einem Druck von 500 atm oder bei 200–300 °C [146]. Das Mineral *Sibirskit* hat die gleiche Zusammensetzung.

Das bei Raumtemperatur beständige Calciummonoborat, CaO · B_2O_3 · 6 H_2O, kristallisiert aus wäßriger Lösung ebenfalls nur, wenn der p_H-Wert der Lösung höher als 11 ist. Die Darstellungsbedingung bei Raumtemperatur und die Kristallstrukturanalyse [147] sprechen dafür, seine Struktur als $Ca[B(OH)_4]_2$ · 2 H_2O zu beschreiben. Aufgrund von Patterson-Projektionen und Darstellung der Elektronenverteilung kam allerdings ein Jahr zuvor ein anderer russischer Autor [148] dazu, es als $Ca[B_2O(OH)_6]$ · 3 H_2O mit einem dem Mineral Pinnoit entsprechenden Anion zu formulieren.

CaO · B_2O_3 · 6 H_2O existiert als stabile Phase im System CaO/B_2O_3/H_2O bei 18 °C, während bei 50 °C CaO · B_2O_3 · 4 H_2O auskristallisiert. Das 4-Hydrat existiert in zwei Kristallformen; die Struktur des monoklinen β-CaB_2O_4 · 4 H_2O wurde untersucht, und die Gitterparameter sowie die Koordination um die B- und Ca-Atome wurden bestimmt. Da Bor nur tetraedrisch koordiniert ist, wird die Formel β-$Ca[B(OH)_4]_2$ vorgeschlagen [149]. Auch die Struktur des rhombisch kristallisierenden α-$Ca[B(OH)_4]_2$ wurde neuerdings [150] untersucht, und es wurden isolierte B–O-Tetraeder nachgewiesen. Die Röntgenpulveraufnahme des Minerals

Frolovit, $CaO \cdot B_2O_3 \cdot 3{,}65\,H_2O$, hat Ähnlichkeit mit der von β-$Ca[B(OH)_4]_2$.

Durch Erhitzen von $CaO \cdot B_2O_3 \cdot 6\,H_2O$ auf 105 °C entsteht das amorphe $CaO \cdot B_2O_3 \cdot 2\,H_2O$, das als primäres Calciummonoborat [144], $\{Ca[BO(OH)_2]_2\}_n$, formuliert wird; bei 150 °C bleibt 1 Mol Wasser erhalten, das erst bei ca. 700 °C abgegeben wird. Beim Versuch, kristallisiertes $CaO \cdot B_2O_3 \cdot 2\,H_2O$ unter hydrothermalen Bedingungen zwischen 100 °C und 190 °C zu erhalten, wurde ein Gemisch aus $2\,CaO \cdot B_2O_3 \cdot H_2O$, dem Mineral *Pandermit* und einem amorphen Produkt erzeugt [146]. Dagegen wurde ein Halbhydrat, $CaO \cdot B_2O_3 \cdot 0{,}5\,H_2O$, beim Erhitzen einer wäßrigen Lösung von CaO und B_2O_3 im Molverhältnis 1:1 auf 200—300 °C erhalten.

3.7.2. Wasserhaltige Calciumtriborate

Die Calcium-(2:3)-Borate bilden eine verwandte Reihe, wie systematische Kristallstrukturuntersuchungen von Clark und Christ [151] zeigen. Im Mineral *Inyoit*, $2\,CaO \cdot 3\,B_2O_3 \cdot 13\,H_2O$ [152], im synthetischen $2\,CaO \cdot 3\,B_2O_3 \cdot 9\,H_2O$ [153], das in ca. einem Monat bei 50 °C in Wasser auf Ulexit-Bruchstücken wächst, und im Mineral *Meyerhofferit* [154], $2\,CaO \cdot 3\,B_2O_3 \cdot 7\,H_2O$, liegt danach das Inselpolyboration $[B_3O_3(OH)_5]^{2-}$ vor. Das bestätigt auch eine genaue Untersuchung des IR-Spektrums von Meyerhofferit [155]. Inyoit läßt sich also als $Ca(H_2O)_4[B_3O_3(OH)_5]$, das synthetische Produkt als $Ca(H_2O)_2[B_3O_3(OH)_5]$ und Meyerhofferit als $Ca(H_2O)[B_3O_3(OH)_5]$ formulieren.

Dagegen liegt im wasserärmeren Mineral *Colemanit* [156], $2\,CaO \cdot 3\,B_2O_3 \cdot 5\,H_2O$, das Kettenpolyboration $\{B_3O_4(OH)_3\}_n^{2n-}$ vor. $\{Ca(H_2O)[B_3O_4(OH)_3]\}_n$ bildet unendliche Ketten längs der kristallographischen a-Achse, vernetzt durch Ionenbindungen über die Ca^{2+}-Ionen zu Schichten. Diese Schichten wiederum sind durch ein System von Wasserstoffbrückenbindungen vernetzt, wobei die OH-Gruppen des ringförmigen Polyanions und die Wassermoleküle benutzt werden. An Colemanit-Einkristallen, die bei unterhalb -2 °C ferroelektrisch sind, wurden ^{11}B-NMR-Spektren [157] bei Raumtemperatur und bei -40 °C aufgenommen; dabei wurde die reversible Umwandlung der beiden Formen beobachtet.

Schließlich liegen im synthetischen $2\,CaO \cdot 3\,B_2O_3 \cdot H_2O$ [158], das durch mehrtägiges Erhitzen von Inyoit auf 400 °C unter Druck entsteht, unendliche Schichten der Zusammensetzung $\{B_3O_5(OH)\}_n^{2n-}$ vor, die durch Ca—O-Bindungen zusammengehalten werden.

Bei den Calcium-(1:3)-Boraten existieren zwei monoklin kristallisierende Mineralien, nämlich *Gowerit*, $CaO \cdot 3\,B_2O_3 \cdot 5\,H_2O$, und *Nobleit*, $CaO \cdot 3\,B_2O_3 \cdot 4\,H_2O$, sowie das synthetische $CaO \cdot 3\,B_2O_3 \cdot 1/3\,H_2O$.

Zwischen Nobleit [159] und Gowerit besteht keine einfache Beziehung, denn beide lassen sich nicht durch Hydratisierung oder Dehydratisierung ineinander umwandeln. Die Bildungs- und Existenzbedingungen beider Mineralien sowie ihre Umwandlung unter der Mutterlauge in ca. 40 Tagen in die Verbindung $3\,CaO \cdot 10\,B_2O_3 \cdot 12\,H_2O$ werden beschrieben [160]. Letztere soll dem Röntgenpulverdiagramm nach mit dem Mineral *Ginorit* identisch sein. Gowerit scheint entweder die Konstitution des Kettenpolyborats, $\{Ca(H_2O)_2[B_6O_7(OH)_6]\}_n$ [160], oder des Inselpolyborats, $Ca(H_2O)[B_3O_3(OH)_4]_2$ [161], zu besitzen. Da Nobleit isostrukturell mit dem Mineral *Tunellit*, $SrO \cdot 3\,B_2O_3 \cdot 4\,H_2O$ [162], ist, dessen Struktur bestimmt wurde, muß seine Strukturformel $\{Ca(H_2O)_3[B_6O_9(OH)_2]\}_n$ lauten.

Das synthetische $CaO \cdot 3\,B_2O_3 \cdot 1/3\,H_2O$ [146] wurde unter hydrothermalen Bedingungen bei 400 °C dargestellt; aus seinem Röntgenpulverdiagramm und seinem Verhalten bei der thermischen Analyse läßt sich noch nichts über seine Struktur aussagen.

3.7.3. Weitere wasserhaltige Calciumpolyborate

Die einzigen bekannten Calcium-(4:5)-Borate sind die Mineralien *Pandermit* oder *Priceit*, $4\,CaO \cdot 5\,B_2O_3 \cdot 7\,H_2O$ (oder $6\,CaO \cdot 7\,B_2O_3 \cdot 10\,H_2O$), und *Tertschit*, $4\,CaO \cdot 5\,B_2O_3 \cdot 20\,H_2O$. Ersteres ist in Wasser leicht löslich; sein thermischer Abbau [123] verläuft recht kompliziert. Das IR-Spektrum von Tertschit [163] zeigt, daß dieses Mineral eng mit Colemanit verwandt sein muß. Über die Strukturen beider Mineralien ist noch nichts bekannt; der Zusammensetzung nach könnte in ihnen das Inselpolyboration $[B_5O_6(OH)_7]^{4-}$ vorliegen.

Die Verbindung der Zusammensetzung $2\,CaO \cdot 5\,B_2O_3 \cdot 5\,H_2O$ [164] wurde in zwei Formen durch hydrothermale Reaktion von CaO mit $B(OH)_3$ dargestellt, und zwar die eine bei 190 °C unter 13 atm Druck beim Ca:B-Verhältnis 1:9 und die andere bei 295 °C—362 °C beim Ca:B-Verhältnis 1:20.

Vom Mineral *Ginorit*, $2\,CaO \cdot 7\,B_2O_3 \cdot 8\,H_2O$ [165], wurde bisher nur das Röntgenpulverdiagramm veröffentlicht. Die Verbindung $3\,CaO \cdot 10\,B_2O_3 \cdot 12\,H_2O$ [160], die dem Mineral Ginorit ähnlich sein soll, entsteht, wenn H_3BO_3 und CaO im Ca:B-Verhältnis $\leqslant 1:10$ hydrothermal bei 190 °C umgesetzt werden.

3.7.4. Wasserfreie Calciumborate

Im System CaO/B_2O_3 existieren die kristallinen Verbindungen $3\,CaO \cdot B_2O_3$, $2\,CaO \cdot B_2O_3$, $CaO \cdot B_2O_3$ und $CaO \cdot 2\,B_2O_3$, von denen die IR-Spektren bekannt sind [166]. Die Darstellungsbedingungen und Schmelzpunkte sind von Lehmann u. Mitarb. [144], die Röntgendaten der ersteren Verbindung von Schäfer [145] angegeben; die Struktur des orthorhombischen Calciummetaborats, $CaO \cdot B_2O_3$, wurde von Marezio u. Mitarb. [167] verfeinert; sie enthält BO_3-Gruppen, die, mit benachbarten BO_3-Gruppen über gemeinsame Sauerstoffatome verbunden, endlose Doppelketten parallel zur kristallographischen c-Achse bilden; dieses Kettenborat kann $\{Ca(BO_2)_2\}_n$ formuliert werden. $CaO \cdot 2\,B_2O_3$ enthält wahrscheinlich $B_4O_7^{2-}$-Gruppen.

3.7.5. Wasserhaltige Strontiumborate

Bei den systematischen Untersuchungen [168] im System $SrO-B_2O_3-H_2O$ wurden als Bodenkörper folgende Verbindungen gefunden, die bei den angegebenen Temperaturen existent sind:

$$SrO \cdot B_2O_3 \cdot 4\,H_2O \quad (0{-}100\,^\circ C),$$

$$SrO \cdot B_2O_3 \cdot H_2O \quad (190\,^\circ C),$$

$$2\,SrO \cdot 3\,B_2O_3 \cdot 9\,H_2O \quad (0{-}30\,^\circ C),$$

$$2\,SrO \cdot 3\,B_2O_3 \cdot 6\,{}^2/_3\,H_2O \quad (100\,^\circ C),$$

$$2\,SrO \cdot 3\,B_2O_3 \cdot 5\,H_2O \quad (190\,^\circ C),$$

$$SrO \cdot 3\,B_2O_3 \cdot 5\,H_2O \quad (0{-}100\,^\circ C) \text{ und}$$

$$SrO \cdot 4\,B_2O_3 \cdot 2\,H_2O \quad (190\,^\circ C).$$

Bei diesen Untersuchungen wurde noch auf Konstitutionsbetrachtungen verzichtet.

Strontiummetaborat, $SrO \cdot B_2O_3 \cdot 4\,H_2O$, kristallisiert sowohl monoklin als auch triklin. Das monokline Produkt enthält seiner Kristallstrukturanalyse nach [169] Inselpolyborationen mit vierbindigen Bor; es besitzt die Strukturformel $Sr[B(OH)_4]_2$.

Im synthetischen $SrO \cdot 3\,B_2O_3 \cdot 5\,H_2O$ soll nach IR-Untersuchungen [133] das Tunellitanion $\{B_6O_9(OH)_2\}_n^{2n-}$ nachweisbar sein.

In der Natur kommen außer diesen Verbindungen folgende Mineralien vor: Veatchit, Paraveatchit, Tunellit, Strontium-Colemanit und Strontioginorit.

Die Zusammensetzungen von *Veatchit* [170] und *Paraveatchit* [171] sind gleich; beide haben sehr ähnliche Röntgenpulverdiagramme und können nur durch Einkristallröntgenuntersuchungen unterschieden werden [172]. Die Formel dieser beiden Mineralien war lange Jahre Objekt von Kontroversen zwischen Lehmann u. Mitarb. [173,174], die die Formel $3\,SrO \cdot 8\,B_2O_3 \cdot 5\,H_2O$ angaben, und Clark [172], die für die Formel $SrO \cdot 3\,B_2O_3 \cdot 2\,H_2O$ eintrat, bis Gandymow u.a. kürzlich [175] die Aufklärung der Röntgenstruktur von Paraveatchit und kurz darauf Clark und Christ [176] die Röntgenstrukturanalyse von Veatchit gelangen. Danach besitzen beide Mineralien die Zusammensetzung $4\,SrO \cdot 11\,B_2O_3 \cdot 7\,H_2O$ und die Struktur $\{Sr_2[B_5O_8(OH)]_2 \cdot B(OH)_3 \cdot H_2O\}_n$ mit dem neuen Schichtenboration $\{B_5O_8(OH)\}_n^{2n-}$.

Das Mineral *Tunellit*, $SrO \cdot 3\,B_2O_3 \cdot 4\,H_2O$, hat eine ganz besondere Struktur, wie seine Kristallstrukturanalyse [8] und die quantitative ^{11}B-NMR-Untersuchung [9] zeigen: Die Struktur besteht nämlich aus unendlichen Schichten von Polyborationen der Form $\{B_6O_9(OH)_2\}_n^{2n-}$. Man kommt bei einer derartigen Strukturformel nicht umhin, ein *Sauerstoffatom mit positiver Ladung* anzunehmen, das drei tetraedrisch koordinierte Boratome bindet. Der dreibindige Sauerstoff kompensiert im Sinne des ersten Postulats von Edwards und Ross [4] eine Kationenladung. Die gleiche Bindung soll auch im Mineral *Nobleit* [162] vorliegen. In der Verbindung $2\,SrO \cdot 3\,B_2O_3 \cdot 5\,H_2O$, die Strontium-Colemanit genannt wird, sollen [174] die Borat-Einheiten noch stärker vernetzt sein als im Kettenpolyborat Colemanit. Strontioginorit hat die Zusammensetzung $2\,SrO \cdot 7\,B_2O_3 \cdot 8\,H_2O$; ob es calciumfrei vorkommt, ist nicht ganz sicher [177].

3.7.6. Wasserfreie Strontiumborate

Phasengleichgewichtsuntersuchungen [178] im System SrO/B_2O_3 zeigen die Existenz der Verbindungen $2\,SrO \cdot B_2O_3$, $SrO \cdot B_2O_3$ und $SrO \cdot 2\,B_2O_3$, vermutlich auch die von $SrO \cdot 3\,B_2O_3$. $2\,SrO \cdot B_2O_3$ enthält nach Kristallstrukturuntersuchungen [179] annähernd ebene $B_2O_5^{4-}$-Gruppen. In $SrO \cdot 2\,B_2O_3$ sind nach zwei verschiedenen Kristallstrukturanalysen [180,181], die nur zu unterschiedlichen Raumgruppen kommen, alle Boratome tetraedrisch koordiniert, während ein einzelnes Sauerstoffatom pro Formeleinheit von drei Boratomen umgeben wird. Diese ungewöhnliche Koordination wurde früher schon beim Tunellit [8] und bei B_2O_3 [182] beobachtet. Über die Strukturen von $SrO \cdot B_2O_3$ und $SrO \cdot 3\,B_2O_3$ ist nichts bekannt.

3.7.7. Bariumborate

Die Darstellungsbedingungen für die monoklinen Metaborate $BaO \cdot B_2O_3 \cdot 4\,H_2O$ und $BaO \cdot B_2O_3 \cdot 5\,H_2O$ werden von Wimba u. Mitarb. [183] angegeben. Letzteres kristallisiert aus verdünnteren Lösungen in Gegenwart einer geringeren Konzentration an $BaCl_2$ als ersteres aus. Nach Kristallstrukturanalysen [184,185] ist $BaO \cdot B_2O_3 \cdot 5\,H_2O$ mit isolierten $B(OH)_4^-$-Gruppen aufgebaut und als $Ba[B(OH)_4]_2 \cdot H_2O$, ersteres dagegen [186] kristallwasserfrei als $Ba[B(OH)_4]_2$ zu formulieren. Diese Verbindung gibt bei 140 °C ihr gesamtes Wasser ab. Bei Siedetemperatur,

$p_H > 11,4$ und einem Ba:B-Verhältnis von größer als 0,5 erhält man als Bodenkörper $BaO \cdot B_2O_3 \cdot 1,67\ H_2O$, das beim Erhitzen auf 300 °C in die Tieftemperaturform von $BaO \cdot B_2O_3$ übergeht [187]. Bei den Bariumtriboraten sollen in kristallisierter Form $2\ BaO \cdot 3\ B_2O_3 \cdot 6\ H_2O$, $BaO \cdot 3\ B_2O_3 \cdot 5\ H_2O$ und $BaO \cdot 3\ B_2O_3 \cdot 4\ H_2O$ existieren, über deren Strukturen aber noch keine Aussagen möglich sind.

Von den wasserfreien Bariumboraten kennt man $3\ BaO \cdot B_2O_3$, $BaO \cdot B_2O_3$, $BaO \cdot 2\ B_2O_3$ und $BaO \cdot 4\ B_2O_3$. In zwei polymorphen Formen existiert $BaO \cdot B_2O_3$ [188]; die rhomboedrische Hochtemperaturform enthält nach ihrer Kristallstrukturbestimmung fast ebene $B_3O_6^{3-}$-Gruppen mit dreibindigem Bor [166,189], die Tieftemperaturform nach ihren IR-Spektren drei- und vierbindiges Bor [187].

Nach der Kristallstrukturanalyse von $BaO \cdot 2\ B_2O_3$ [190] besteht dieses abwechselnd aus Penta- und Triborat-Ringsystemen, die über Sauerstoffbrücken ein dreidimensionales Netzwerk bilden. In den Kanälen dieses Netzwerkes liegen die Ba^{2+}-Ionen; die Strukturformel ist am besten mit $\{Ba_2B_8O_{14}\}_n$ anzugeben.

3.8. Andere Borate mit Metallkationen

Weitere wasserhaltige Borate gibt es mit Ag^+, Tl^+, Ni^{2+}, Co^{2+}, Zn^{2+}, Fe^{2+}, Mn^{2+}, Cd^{2+}, Sn^{2+}, Pb^{2+}, Al^{3+} und Cr^{3+}. Über deren Strukturen ist noch sehr wenig bekannt; lediglich von der Verbindung $2\ ZnO \cdot 3\ B_2O_3 \cdot 7\ H_2O$ scheint die Strukturformel $Zn(H_2O)\ [B_3O_3(OH)_5]$ bestätigt zu sein [191].

Weitere wasserfreie Borate gibt es mit

Ag^+, Tl^+, Ni^{2+}, Co^{2+}, Zn^{2+}, Fe^{2+}, Mn^{2+}, Cu^{2+}, Cd^{2+}, Hg^{2+}, Sn^{2+}, Pb^{2+}, Al^{3+}, Ga^{3+}, Cr^{3+}, V^{3+}, Fe^{3+}, Ti^{3+}, Bi^{3+}, Sc^{3+}, In^{3+}, Lu^{3+}, Y^{3+}, Sm^{3+}, Nd^{3+}, Ce^{3+}, La^{3+}, Ti^{4+}, Sn^{4+} (in Verbindung mit zweiwertigen Metallionen), Th^{4+}, V^{5+} (in Verbindung mit anderen Metallionen) und Ta^{5+}.

Die Verbindungen des Typs $Me^{3+}BO_3^{3-}$ mit dem monomeren BO_3^{3-}-Ion, können — entsprechend dem CO_3^{3-}-Ion — Aragonit-, Calcit- oder Vaterit-Struktur [192] besitzen. Die Verbindungen $Me^{2+}Sn(BO_3)_2$ mit Me=Mg, Ni, Co, Mn, Cd, Sr und Ba haben Dolomit-Struktur. Ebenfalls isolierte BO_3^{3-}-Gruppen neben zu Ketten verknüpften AlO_4-Tetraedern und sechsfach koordinierten Ca^{2+}-Ionen enthält die Struktur von $CaAlBO_4$ [193]. BPO_4, $BAsO_4$ (und $BSbO_4$) besitzen eine Struktur [194], die sich als tetragonale Deformation einer SiO_2-Modifikation, der β-Form des Cristobalits, darstellen läßt. Dagegen wurden die vermuteten tetraedrischen BO_4^{5-}-Ionen röntgenographisch in der Verbindung $Fe_3^{III}O_2(BO_4)$ nachgewiesen, die mit dem Mineral *Norbergit*, $Mg_2SiO_4 \cdot Mg(OH,F)_2$, isostrukturell ist [195]. Ein neuer Strukturtyp findet sich auch in der Verbindung $4\,ZnO \cdot 3\,B_2O_3$ [196]), der dem des dreidimensionalen Gerüstes des $Al_3Si_3O_{12}^{3-}$-Ions im Mineral *Sodalith* sehr ähnlich sein soll. Er besteht aus großen $B_6O_{12}^{6-}$-Ringen mit den Sauerstoffatomen auf regulären Tetraederecken, in deren Zentren sich die Boratome befinden; die Tetraeder sind miteinander zum Ring verbunden. $Hg_4O[B_6O_{12}]$ hat die gleiche Struktur [197].

3.9. Polyborate mit organischen Kationen

3.9.1. Bisher bekannte Polyborate mit organischen Kationen

Polyborate mit organischen Kationen wurden bisher nur aus wäßrigen Lösungen gewonnen. In einer Arbeit aus dem Jahre 1921 [105] wird die Darstellung von zwei Guanidiniumboraten (Gu = Guanidinium) beschrieben, und zwar $Gu_2O \cdot 2\,B_2O_3 \cdot 4\,H_2O$ durch Kochen von Guanidiniumcarbonat mit Borsäure in Wasser, und $Gu_2O \cdot 5\,B_2O_3 \cdot 8\,H_2O$ aus kalten konzentrierten Lösungen von Borax, Borsäure und Guanidiniumchlorid.

Folgende Alkylammoniumborate wurden aus Borsäure und überschüssigem Alkylamin erhalten [198]:

$$CH_3NH_3B_5O_8 \cdot 4\,H_2O,$$

$$(CH_3)_2NH_2B_5O_8 \cdot 2\,H_2O,$$

$$(CH_3)_3NHB_5O_8 \cdot 2\,H_2O,$$

$$C_2H_5NH_3B_5O_8 \cdot 2\,H_2O,$$

$$(C_2H_5)_2NH_2B_5O_8 \cdot 2\,H_2O \text{ und}$$

$$(C_2H_5)_3NHB_5O_8 \cdot 2\,H_2O.$$

Kocht man Borsäure in Dimethylformamid (DMF) [199], so zersetzt sich das Lösungsmittel zu Dimethylamin und Ameisensäure, und es kristallisiert aus der Lösung $(CH_3)_2NH_2B_5O_8 \cdot 2\,H_2O \cdot 1\,DMF$ aus, das auch beim Einleiten von Dimethylamin in eine siedende Lösung von Borsäure in DMF entsteht. Läßt man in wäßriger Lösung Borsäure mit Formamid $(HCONH_2)$ [200] reagieren, so hydrolysiert Formamid zu NH_3 und Ameisensäure, und aus der Lösung kristallisiert $NH_4B_5O_8 \cdot 4\,H_2O \cdot HCONH_2$ aus.

Aus wäßriger Lösung gewonnene Polyborate quaternärer Ammoniumbasen [201] besitzen nach dem Trocknen im Vakuum bei 55—90 °C entweder die Formeln

$QH_4B_5O_{10}$ (Q = Tetramethylammonium, Tetraäthylammonium, Trimethylanilinium, Allyldimethylanilinium, Trimethylfluorenylammonium, Triäthylfluorenylammonium oder Benzyltrimethylammonium),

$QH_4B_5O_{10} \cdot n\, H_2O$ ($n = 1{:}Q$ = Tetraallylammonium, Triallyl-n-propylammonium, $n = 3{:}Q$ = Trimethyl-1-acenaphthenylammonium, Benzyltriäthylammonium und Benzyl-tri-n-butylammonium) oder
$QH_4B_5O_{10} \cdot H_3BO_3$ (Q = Cinnamyltriäthylammonium, Benzyldimethylallylammonium und Benzyldimethyl-n-propylammonium).

Über die Strukturen dieser Polyborate ist wenig bekannt. Die Summenformeln lassen vermuten, daß es sich bei allen Polyboraten mit organischen Kationen mit Ausnahme des Guanidiniumborats $Gu_2O \cdot 2\, B_2O_3 \cdot 4\, H_2O$ um Pentoborate handelt. Bei den Verbindungen $QH_4B_5O_{10} \cdot H_3BO_3$ ist es fraglich, ob das pauschal als Kristallborsäure in der Formel angegebene H_3BO_3 in Wirklichkeit als solches vorliegt. Deshalb wurde ein Vertreter dieser Gruppe nochmals aus wäßriger Lösung dargestellt und genauer untersucht.

Die Aufstellung von Strukturmodellen scheitert bei den aus wäßriger Lösung gewonnenen Polyboraten meist daran, daß man nicht weiß, ob das in ihnen enthaltene Wasser als Kristallwasser oder als Strukturwasser vorliegt. Aus der Zusammensetzung der durch Esterhydrolyse entstandenen Verbindungen kann man aber Strukturhinweise erhalten, und daher wurden diese Untersuchungen auch auf die Darstellung von Polyboraten mit organischen Kationen ausgedehnt [202].

3.9.2. Guanidiniumpolyborate

Bei der Hydrolyse einer Lösung von Borsäuretrimethylester in Testbenzin fallen in Gegenwart von Guanidin in Äthanol Gemische von Polyboraten aus, die bei verschiedenen Konzentrationsverhältnissen drei Polyborate mit den Base:Säure-Verhältnissen $2{:}3$, $3{:}4$ und $4{:}5$ ergeben. Diese drei relativ basereichen Guanidiniumborate enthalten nach dem Trocknen kaum noch Lösungsmittel adsorbiert, wie man es bei anderen, durch Esterhydrolyse gewonnenen Polyboraten findet. Dagegen muß in diesen Salzen eine stickstoff-haltige Verbindung adsorbiert sein, die keine starke Base ist; das zeigt ein Vergleich der Stickstoff-Werte, die man einmal durch Titration der freigesetzten Base, zum anderen mikroanalytisch mit Hilfe des CHN-Modells 185 oder durch Aufschluß und Destillation nach *Kjeldahl* erhält. Die Analysenergebnisse stimmen recht gut, wenn man annimmt, daß Harnstoff adsorbiert ist, der nach folgender Gleichung durch Hydrolyse von Guanidin entstanden ist:

$$
\begin{array}{c}
\underset{\diagdown NH_2}{\overset{\diagup NH_2}{C{=}NH}} + H_2O \longrightarrow \underset{\diagdown NH_2}{\overset{\diagup NH_2}{C{=}O}} + NH_3 .
\end{array}
$$

Beim thermischen Abbau eines Salzes mit dem Base:Säure-Verhältnis $2{:}3$ erleidet die Substanz bis 67 °C praktisch keinen Gewichtsverlust, bis 130 °C sind 22%, bis 137 °C 38%, bis 215 °C 50% und bis 220 °C 55% abgebaut. Danach verläuft der Abbau nahezu stufenlos, bis bei 550 °C 76% der Substanz zersetzt sind. Der Rückstand besitzt ein Base:Säure-Verhältnis von 1:6,9.

Die drei Guanidiniumborate unterscheiden sich sowohl in ihren IR-Spektren (Tabelle 4) als auch in den Röntgenpulverdiagrammen (Tabelle 5).

Gegenüber diesen drei durch Esterhydrolyse dargestellten Guanidiniumpolyboraten wurde aus wäßriger Lösung nur *ein Guanidiniumpolyborat* gewonnen. Beim Nacharbeiten der älteren Vorschrift [105) zur Darstellung der Verbindung $Gu_2O \cdot 2\,B_2O_3 \cdot 4\,H_2O$ wurde nach mehrmaligem Umkristallisieren aus Wasser ein Guanidiniumborat erhalten, das schon bei Raumtemperatur über P_2O_5 die stöchiometrische Zusammensetzung $Gu_2O \cdot 5\,B_2O_3 \cdot 4\,H_2O$ erreicht und sich auch beim Trocknen im Ölpumpenvakuum über P_2O_5 bei 60 °C analytisch nicht mehr verändert. Wahrscheinlich ist es hier ähnlich wie bei den Ammoniumpentaboraten, bei denen das aus wäßriger Lösung gewonnene $(NH_4)_2O \cdot 5\,B_2O_3 \cdot 8\,H_2O$ und das kristallwasserfreie Mineral Lardellerit $(NH_4)_2O \cdot 5\,B_2O_3 \cdot 4\,H_2O$ auch leicht und reversibel ineinander übergehen. In Analogie zu diesem Beispiel kann man dem entwässerten Guanidiniumpentaborat die Strukturformel $Gu[B_5O_6(OH)_4]$ zuschreiben.

3.9.3. Piperidiniumpolyborate

Nimmt man Piperidin als Base bei der Hydrolyse von Borsäuretrimethylester, so fällt zunächst kein Niederschlag aus. Erst beim Einengen der Lösung im Vakuum erhält man Gemische von Piperidiniumboraten mit den Base:Säure-Verhältnissen 1:4 bis 1:5. Durch Umkristallisieren aus entwässertem Aceton kann man beide Polyborate rein isolieren. Die IR-Spektren des Tetra- und Pentaborats (Tabelle 4) unterscheiden sich in Lage und Intensität einiger Banden. Beide Piperidiniumborate lösen sich gut in Methanol. Die Bildung von 1:4- und 1:5-Boraten findet eine Parallele bei der Esterhydrolyse in Gegenwart von Ammoniak, und es ist anzunehmen, daß die Piperidiniumpolyborate ähnliche Strukturen wie die entsprechenden Ammoniumpolyborate haben.

3.9.4. Tetraalkylammoniumpolyborate

Mit einer Tetramethylammoniumhydroxid-Lösung in Isopropanol/Methanol fällt bei der Hydrolyse einer Lösung des Borsäuretrimethylesters sofort, auch bei starkem Variieren der Konzentrationsverhältnisse und

Tabelle 4. IR-Spektren (cm⁻¹) der dargestellten Polyborate

Guanidinium-				Piperidinium-		Tetramethylammonium-		Zuordnung
(2:3)-Borat	(3:4)-Borat	(4:5)-Borat	(1:5)-Borat*	(1:5)-Borat	(1:4)-Borat	(1:5)-Borat	(1:5)-Borat · aq*	
684 w, b	669 w	675 b	675 sh	675 sh	675 sh	681 w	660 vw	δ (Ring)
714 w, sh		710 m	711 s	708 s	710 s	712 vs	710 s	
				725 m, sh	726 m, sh	728 s	725 m	
							742 w	
768 m	770 m	754 w, sh	750 w, sh	751 w		752 w, sh	774 vs	δ (B_3—O—B_3)
		773 s	780 s	781 s	781 m	775 vs	783 w, sh	
						784 w, sh	796 w	
844 m, b	842 m, b	810 m, sh		825 m	817 s		807 w, sh	
		840 vs	875 vw	869 m	865 s	848 s		ν (O—B_4)
882 s	881 b, sh	882 s, sh	925 vs	926 vs	929 s	924 vs	898 s	
				947 sh	934 sh	952 sh	921 vs	
1020 s, b	1025 s, b	1030 s, b	1032 s, b	1029 s, b	1031 m	1029 s	1009 s	ν (B_3—O—B_4)
1055 s	1057 s	1055 m	1055 sh	1064 m	1070 sh	1058 m	1024 sh	
				1086 m	1085 m	1080 sh	1068 s	
		1115 m, b		1118 m	1120 w, b	1107 m	1089 sh	
							1124 b	

$\nu(B_4\text{–}O\text{–}B_4)$, $\nu(B_3\text{–}O\text{–}B_3)$	$\nu(O\text{–}B_3)$	$\delta(OH)$	$\nu(CH, CH_2)$	$\nu(OH)$
1180 s; 1202 w	1310 s; 1390 m; 1414 s; 1468 w	1496 m	2974 w; 3000 w	3316 m; 3522 m
1160 m, b	1318 s, b; 1410 s, b; 1458 sh	1494 m	2995 sh	3278 s; 3390 m
1281 sh	1325 m, b; 1391 s, b; 1430 s, b; 1464 sh	1588 m; 1626 s	2960 vs	3410 s, b
1145 m; 1289 sh	1323 m; 1396 w; 1431 s, b; 1464 sh	1587 w; 1625 m	2965 s	3240 b; 3395 s
1287 sh	1335 m, b; 1415 s, b; 1452 sh	1568 sh; 1595 sh; 1670 s, b		3210 s, b; 3383 m; 3445 m
1176 w; 1287 sh	1340 sh; 1415 s, b; 1446 sh	1575 sh; 1600 sh; 1663 s, b		3200 s, b; 3378 vs; 3420 s, b
1166 sh	1328 b; 1390 s, b; 1407 b; 1448 sh	1569 sh; 1634 sh; 1674 s, b		3200 sh; 3360 s, b; 3420 s, b
1160 sh	1337 w; 1391 s, b; 1413 b	1594 sh; 1675 s, b		3183 s, b; 3377 s, b

(* Aus Wasser gewonnene Borate; b=breit, m=mittel, mw=mittelschwach, s=stark, sh=Schulter, vs=sehr stark, vw=sehr schwach, w=schwach)

Tabelle 4 (Fortsetzung)

Tetraäthylammonium- (1:5)-Borat	Tetraäthylammonium- (1:5)-Borat*	Tetrapropylammonium- (1:5)—(1:7)-Borat	Tetrapropylammonium- (1:7)-Borat*	Tetrabutylammonium- (1:7)-Borat	Tetrabutylammonium- (1:7)-Borat*	Tributylammonium- (1:7)-Borat	Tributylammonium- (1:7)-Borat*	Zuordnung
662 w	660 w	675 m	673 s	673 m	676 m	675 s	674 s	δ (Ring)
710 s	709 s	709 s	706 m	705 m	710 m	706 m	706 m	
726 m	725 m	726 w	723 w		727 w	722 w	721 w	
744 w	741 w		745 w, sh	741 w	740 w	743 w	741 w	δ (B_3—O—B_3)
774 vs	772 vs	769 sh	765 sh					
		784 s	782 s	784 s	781 s	786 s	785 s	
797 w	795 w							
	808 w, sh							
		819 sh	815 w	815 sh				
						831 m	831 m	
		842 b	840 w		842 w, sh			
						873 w	873 w	
			876 m	876 sh		878 m	879 m	
900 s	898 s							ν (O—B_4)
920 s	920 s	926 vs	923 vs	925 vs	921 vs	929 vs	928 vs	
			940 sh	945 sh		945 sh	943 sh	
		973 s	970 s					
1008 s	1007 s				1011 sh			
1027 sh	1025 sh	1032 s	1035 s	1030 vs	1026 s	1032 vs	1028 s	ν (B_3—O—B_4)
1066 s	1067 s		1060 w	1057 sh	1045 w	1043 sh	1044 s	
1091 sh	1088 w, sh							
1119 b	1125 b	1107 s, b		1106 s	1111 b	1108 s	1104 w	

								Zuordnung
1180 s	1179 s	1148 sh	1150 s		1178 sh	1144 sh / 1177 sh	1147 m / 1179 sh	ν (B$_4$—O—B$_4$)
1201 b	1200 b	1234 sh / 1261 sh	1229 m / 1257 sh	1224 sh / 1258 sh		1230 sh / 1256 w	1227 m	(B$_3\nu$—O—B$_3$)
1311 s,b / 1390 w,b / 1411 s,b / 1470 sh	1310 s / 1386 w / 1412 s / 1465 w	1318 w,b / 1385 w,b / 1436 s	1310 s / 1428 s	1310 sh / 1381 b / 1430 s	1305 s / 1390 sh / 1425 s,b	1320 w / 1386 b / 1432 s	1313 s / 1425 b	ν (O—B$_3$)
1500 m,b	1495 m	1491 sh / 1651 m,b	1489 w	1652 b	1650 b	1649 m	1625 b	δ (OH)
2980 w / 3005 m	2977 w / 3002 w	2980 s / 3000 w	2977 w / 2995 w	2977 vs	2965 s	2975 s	2980 s	ν (CH, CH$_2$)
3338 m / 3540 m	3320 m / 3530 m	3390 s / 3445 sh	3355 s	3395 s / 3445 sh	3375 s	3400 s / 3440 sh	3405 s	ν (OH)

(* Aus Wasser gewonnene Borate; b = breit, m = mittel, mw = mittelschwach, s = stark, sh = Schulter, vs = sehr stark, vw = sehr schwach, w = schwach)

255

Tabelle 5. *Röntgendaten der dargestellten Polyborate* (Θ in °, Cu K_α-Strahlung, Ni-Filter)

2θ	Guanidinium- (3:4)-Borat	Guanidinium- (4:5)-Borat	Guanidinium- (1:5)-Borat*	Piperidinium- (1:5)-Borat	Tetramethyl-ammonium- (1:5)-Borat	Tetramethyl-ammonium- (1:5)-Borat·aq*	Tetraäthyl-ammonium- (1:5)-Borat	Tetraäthyl-ammonium- (1:5)-Borat*	Tetrapropyl-ammonium- (1:7)-Borat	Tetrapropyl-ammonium- (1:7)-Borat*	Tetrabutyl-ammonium- (1:7)-Borat	Tributyl-ammonium- (1:7)-Borat	Tributyl-ammonium- (1:7)-Borat*
3.0		3.0 vw											
3.2							3.2 vw	3.2 vw					
3.4	3.4 m								3.4 vw	3.4 w			
3.9	3.9 s												
4.5												4.5 m	4.5 m
4.6									4.6 vw				
4.7				4.7 vw									
4.8						4.8 mw		4.8 mw			4.8 vs		
4.9							4.9						
5.0						5.0 w		5.0 mw		5.0 mw	5.0 vs		
5.1							5.1 mw			5.1 vs		5.1 vs	5.1 vs
5.2	5.2 vw								5.2 vs				
5.4				5.4 mw							5.4 vw	5.4 mw	5.4 mw
5.5		5.5 w			5.5 s								
5.6		5.6 vw		5.6 mw									
6.0		6.0 vw				6.0 w	6.0 w	6.0 mw				6.0 vw	6.0 vw
6.1										6.1 vw			
6.2									6.2 mw				
6.3											6.3 mw		
6.4						6.4 vw	6.4 vw	6.4 vw					
6.5	6.5 vw												6.5 w
6.6									6.6 m	6.6 m	6.6 mw	6.6 w	
6.8										6.8 s			
6.9					6.9 m	6.9 mw	6.9 mw	6.9 mw	6.9 mw		6.9 mw		
7.0													7.0 vw
7.1					7.1 w	7.1 w		7.1 w				7.1 mw	
7.2	7.2 s			7.2 m			7.2 w						7.2 mw
7.3										7.3 vw		7.3 m	
7.4									7.4 w		7.4 vw		
7.5	7.5 m	7.5 s							7.5 mw	7.5 vw	7.5 w		
7.6					7.6 w	7.6 w	7.6 w	7.6 w					
7.7												7.7 mw	7.7 w
7.8	7.8 m												
7.9											7.9 vw		
8.0		8.0 mw											
8.1						8.1 s	8.1 vs	8.1 vs		8.1 m, b			
8.2					8.2 vw				8.2 ms				
8.6									8.6 vw		8.6 mw		
8.7												8.7 w	8.7 vw
8.8		8.8 mw		8.8 m	8.8 s				8.8 w	8.8 w			
9.0										9.0 vw			
9.1					9.1 w	9.1 vs	9.1 vs	9.1 vs	9.1 vw			9.1 vw	9.1 vw
9.3	9.3 m												
9.4				9.4 m									
9.6								9.6 vs			9.6 s		
9.7							9.7 ms					9.7 s	9.7 s
9.8										9.8 ms			
9.9	9.9 w								9.9 s		9.9 vs		
10.0					10.0 w	10.0 mw	10.0 w	10.0 mw					
10.1		10.1 vw		10.1 s									
10.2									10.2 m	10.2 m			
10.3												10.3 vs	10.3 vs
10.4					10.4 m				10.4 ms	10.4 m	10.4 m		
10.6					10.6 m						10.6 m	10.6 mw	
10.7									10.7 m	10.7 m, b			

11.2 vs	10.9 vw	11.1 w, b	11.5 m	10.8 w	10.9 w	10.8 w	10.9 m	11.3 vw	11.2 mw	11.1 m	11.1 m
11.8 w	11.8 vs	11.7 m	12.5 s	11.2 m	11.3 m	11.2 m	11.3 vw	11.6 m	11.4 w	11.6 ms	11.6 ms
12.4 m	12.1 s	12.0 w	14.8 b	11.8 m	11.9 mw	11.7 mw	11.7 s	12.1 w	11.7 vw	12.0 m	12.0 m
13.0 vw b	12.5 m	12.6 vs	15.0 b	12.1 vs	12.1 s	12.0 m	12.1 mw	12.5 m	12.3 m	12.3 vw	12.7 w
14.7 m	13.4 vw	13.3 w	15.5 mw	12.9 w	12.9 w	12.9 mw	12.5 ms	12.8 mw	12.8 m	12.7 w	13.2 ms
15.1 w	14.8 vw	13.7 w	16.5 w	13.2 w	13.4 vw	14.2 mw	12.9 ms	13.3 mw	14.0 vw	13.2 ms	15.6 m
15.8 w	15.5 w	14.5 vw	16.6 w	13.5 vw	13.6 vw	14.5 m	13.3 mw	14.2 mw	14.5 vw	13.5 vw	16.5 w
17.7 w	15.9 vw	15.1 vw	18.3 vw	14.2 m	14.4 w	14.7 w	13.7 vw	15.2 vw	15.1 w	15.6 m	18.4 w
	16.9 w	15.7 mw		14.5 s	14.5 w	14.9 vw	14.2 mw	16.1 mw	15.9 mw	16.5 w	19.0 w
	17.4 w	15.8 mw		14.9 m	14.7 w	15.9 vw	15.2 w	16.6 vw	19.4 vw	18.4 w	
		16.8 m		15.9 vw	15.0 mw	16.1 vw	15.7 vw	19.8 mw		19.1 w	
		17.8 vw		16.1 w	15.6 vw	17.3 w	16.6 mw			20.1 w	
		21.8 vw		16.9 w	16.1 w	18.2 vw	17.2 vw				
				17.4 w	17.4 vw		20.0 vw				
				19.2 w	18.3 vw		21.2 w				
				20.3 w	19.2 vw						
					20.3 vw						

(* Aus Wasser gewonnene Borate; Intensitäten geschätzt: b = breit, m = mittel, ms = mittelstark, mw = mittelschwach, s = stark, vs = sehr stark, vw = sehr schwach, w = schwach)

257

der zugesetzten Wassermenge, ein Salz der Zusammensetzung 1:5 aus. Lediglich bei zu geringem Wassergehalt oder zu hoher Boresterkonzentration ist keine Fällung zu erzielen. Die Analysenwerte des Salzes entsprechen schon beim Aufbewahren bei Raumtemperatur im Vakuumexsikkator über P_2O_5 der Formel $(CH_3)_4NB_5O_8 \cdot 2\,H_2O$ und ändern sich auch beim Trocknen im Vakuum von 10 Torr bei 50 °C über P_2O_5 nicht mehr. Dagegen enthält das durch Eindampfen im Vakuum aus wäßriger Lösung gewonnene und über P_2O_5 bei Raumtemperatur getrocknete Tetramethylammonium-(1:5)-Borat pro Formeleinheit noch 4 Mole Wasser mehr, hat also die Zusammensetzung $(CH_3)_4NB_5O_8 \cdot 6\,H_2O$. Dieses Wasser wird unter Bildung des 2-Hydrats beim Trocknen im Vakuum bis 50 °C abgegeben. Peterson u. Mitarb. [201] wollen allerdings schon beim Trocknen im Vakuum bei Raumtemperatur ein Salz der Formel $(CH_3)_4NH_4B_5O_{10} = (CH_3)_4NB_5O_8 \cdot 2\,H_2O$ erhalten haben. Die aus Wasser gewonnene wasserreichere Verbindung unterscheidet sich auch sowohl in den IR-Spektren als auch in den Röntgendaten von dem durch Boresterhydrolyse erhaltenen Produkt (Tabellen 4 und 5).

Bei der Esterhydrolyse in Anwesenheit von Tetraäthylammoniumhydroxid in Methanol ergeben sich, weitgehend unabhängig von den Ausgangsmengen der Reaktionspartner, beim Eindampfen im Vakuum immer Salze mit dem Base:Säure-Verhältnis 1:5, die in Methanol löslich sind. Auch aus wäßrigem Medium wurde die gleiche Verbindung $(C_2H_5)_4NB_5O_8 \cdot 2\,H_2O$ erhalten und damit das Ergebnis von Peterson u.a. [201] in diesem Falle bestätigt. Dieses Salz hat nahezu das gleiche IR-Spektrum und das gleiche Röntgenpulverdiagramm wie das aus organischen Lösungsmitteln gewonnene Produkt (Tabellen 4 u. 5).

In Gegenwart von Tetra-n-propylammoniumhydroxid erhält man nach dem Waschen mit Aceton Polyborate, in denen das Verhältnis N:B zwischen 1:5 und 1:7 variiert. Kocht man in absolutem Aceton unter Rückfluß und filtriert in der Hitze, so kann man das reine Salz mit dem Base:Säure-Verhältnis 1:7 isolieren. Dieses löst sich auch in Methanol etwas schwerer als das Gemisch. Das IR-Spektrum des Gemisches besitzt Banden (Tabelle 4), die teils im Spektrum des Tetraäthylammonium-(1:5)-Borats, teils in dem des Tetra-n-butylammonium-(1:7)-Borats zu finden sind. Auch aus wäßriger Lösung ist ein 1:7-Borat zu isolieren, dessen IR- und Röntgendaten sich merklich von denen des aus organischem Lösungsmittel erhaltenen Boratgemisches unterscheiden.

In Anwesenheit von Tetra-n-butylammoniumhydroxid führt die Hydrolyse des Boresters im organischen Medium zu Salzen mit dem Base:Säure-Verhältnis 1:7, wenn man vorher bei Raumtemperatur mit Aceton extrahiert hat. Auch aus wäßriger Lösung ist das 1:7-Borat zu gewinnen, dessen IR-Spektrum dem des durch Esterhydrolyse gewonnenen Produkts gleicht.

Organische Kationen neigen offenbar im Gegensatz zu anorganischen dazu, schon aus wäßriger Lösung kristallwasserfreie Polyborate zu bilden. Das Tetraäthyl-, Tetra-n-propyl- und Tetra-n-butylammoniumsalz, die durch Eindampfen im Vakuum aus wäßriger Lösung gewonnen werden, haben analytisch die gleichen Zusammensetzungen wie die durch Esterhydrolyse erhaltenen Salze und sind kristallwasserfrei. Es ist daher anzunehmen, daß

$$[(C_2H_5)_4N] \ B_5O_8 \cdot 2 \ H_2O,$$

$$[(C_3H_7)_4N] \ B_7O_{11} \cdot 5 \ H_2O \ \text{und}$$

$$[(C_4H_9)_4N] \ B_7O_{11} \cdot 5 \ H_2O$$

die Strukturformeln

$$[(C_2H_5)_4N] \ [B_5O_6(OH)_4],$$

$$[(C_3H_7)_4N] \ [B_7O_6(OH)_{10}] \ \text{und}$$

$$[(C_4H_9)_4N] \ [B_7O_6(OH)_{10}]$$

haben. Lediglich das Tetramethylammoniumborat scheint aus Wasser mit Kristallwasser zu kristallisieren; es besitzt die Zusammensetzung $[(CH_3)_4N] \ B_5O_8 \cdot 6 \ H_2O$, während das durch Boresterhydrolyse erhaltene Produkt $[(CH_3)_4N]B_5O_8 \cdot 2 \ H_2O$ ist. Ersteres ist demnach als $[(CH_3)_4N]$ $[B_5O_6(OH)_4] \cdot 4 \ H_2O$, letzteres als $[(CH_3)_4N] \ [B_5O_6(OH)_4]$ zu formulieren.

Um diese Strukturformeln bestätigen zu können, wäre es wünschenswert, die Anionengewichte der Salze zu kennen. Diese Salze lösen sich auch in Wasser und wasserähnlichen organischen Lösungsmitteln. Aus IR-Untersuchungen [28] weiß man, daß stark verdünnte wäßrige Lösungen von Polyboraten die Eigenschaften gemischter Lösungen von Borsäure $B(OH)_3$ und Orthoborat $B(OH)_4^-$ besitzen. In organischen Lösungsmitteln war diese Dissoziation noch nicht festgestellt worden. Daher war es interessant, das Molgewicht der im organischen Lösungsmittel löslichen Polyborate mit organischem Kation festzustellen. Das mit Hilfe der Ultrazentrifuge an einer 5%igen Lösung von Tetra-n-butylammoniumjodid in Methanol ermittelte mittlere Molekulargewicht sinkt bei Zugabe einer 1%igen Lösung von Tetra-n-butylammoniumheptaborat um etwa 7%. Das bedeutet, daß sich beim Lösen aus dem Heptaborat kleinere Teilchen gebildet haben müssen, die auch sedimentieren und das mittlere Molekulargewicht verringern. In einer Lösung in Aceton, worin das Tetra-n-butylammoniumheptaborat auch löslich ist, ergibt sich das gleiche Bild. In einem organischen Lösungsmittel erfolgt also wie in Wasser eine Solvolyse der Polyborationen, in sehr verdünnter Lösung sogar ein Abbau zu monomeren Teilchen.

Wie Tabelle 5 zeigt, sind die isolierten festen Polyborate röntgenkristallin. Mehr zu sagen ist nicht möglich, da Isotypie mit Polyboraten bekannter Strukturen, vermutlich wegen der großen Kationen, nicht festgestellt wurde. Die IR-Spektren (Tabelle 4) sind so komplex, daß nur aus dem Nichtvorliegen gewisser Banden einige B—O-Bindungen ausgeschlossen werden können (Zuordnungen der Banden vor allem nach 11,203,204)).

Die erstmals dargestellten *1:7-Borate* werden nur mit Kationen erhalten, die einen großen Raumbedarf haben. Während mit n-Propylammoniumhydroxid noch ein Gemisch des 1:7-Borats mit dem 1:5-Borat anfällt, wird mit n-Butylammoniumhydroxid nur noch ein 1:7-Borat erhalten. Für die 1:7-Borate, deren Anionen wahrscheinlich die Formel $[B_7O_6(OH)_{10}]^-$ besitzen, kann man über die mögliche Struktur kaum Angaben machen. Die 1:7-Borate neigen dazu, beim Trocknen im Vakuum bei erhöhter Temperatur das Base:Säure-Verhältnis 1:6,5 anzunehmen; die Analysenwerte entsprechen dann der Anionenformel $[B_{13}O_{11}(OH)_{19}]^{2-}$. Der hohe Gehalt an OH-Gruppen in diesen beiden Anionen könnte zu der Vermutung Anlaß geben, daß in ihnen jeweils durch H-Brücken gebundene Borsäure vorliegt. In diesem Zusammenhang muß noch einmal ins Gedächtnis gerufen werden, daß Peterson u. Mitarb. [201] die bei ihren Untersuchungen mit quaternären Ammoniumbasen aus wäßriger Lösung erhaltenen Polyborate mit dem Base:Säure-Verhältnis 1:6 als $QH_4B_5O_{10} \cdot H_3BO_3$ formulieren und aufgrund des Verhaltens der Salze beim Erhitzen das Vorliegen von HBO_2-Gruppen ausschließen. Danach wäre es denkbar, daß in den 1:7-Boraten aufgrund der sperrigen Kationen Hohlräume entstehen, in die sich H_3BO_3 einlagert.

3.9.5. Benzyldimethyl-n-propylammoniumpolyborate

Aus diesem Grunde wurde versucht, eines der von Peterson [201] beschriebenen „Hexaborate", nämlich das mit Benzyldimethyl-n-propylammonium als Kation, noch einmal darzustellen, um zu prüfen, ob es tatsächlich H_3BO_3 enthält. Beim Eindampfen einer wäßrigen Lösung der quaternären Ammoniumbase mit überschüssiger Borsäure im Rotationsvakuumverdampfer fällt zunächst Borsäure aus. Dann erhält man ein Polyborat, das tatsächlich ein Base:Säure-Verhältnis von etwa 1:6,3 hat. Kocht man dieses Produkt in absol. Methanol unter Rückfluß, so scheidet sich ein Polyborat mit dem Base:Säure-Verhältnis von annähernd 1:7 aus. Beim Einengen des Filtrats und Kühlen erhält man zunächst in geringer Menge ein stark klebriges, wahrscheinlich polymeres röntgenamorphes Produkt, das ein Base:Säure-Verhältnis von etwa 1:2 besitzt, und dann ein Polyborat mit dem Base:Säure-Verhältnis 1:5.

Tabelle 6

IR-Daten (Reflexe in cm^{-1})			Röntgen-Daten (2 ϑ in °)		
Benzyldimethyl-n-propyl-ammonium-		Benzyl-dimethyl-ammonium-	Benzyldimethyl-n-propyl-ammonium-		Benzyl-dimethyl-ammonium-
(1:5)-Borat*	(1:7)-Borat*	(1:7)-Borat*	(1:5)-Borat*	(1:7)-Borat*	(1:7)-Borat*
		652 s	9.1 mw		9.2 vs
660 sh		662 sh	10.5 mw	10.5 w	10.3 w
674 m	675 m	681 sh	10.9 w	10.9 w	
704 s	704 s	704 s, b			
720 sh	721 sh	724 m	11.9 mw	11.9 vw	
728 m	735 m	749 s	12.4 mw	12.4 w	
	768 sh	775 vs	13.2 vw		
778 vs	783 vs	782 sh			13.7 m
840 w	844 m	839 m	14.1 w		14.0 mw
863 sh	865 sh	856 sh		14.8 w	14.6 mw
874 m	878 w	873 m	15.1 w	15.0 mw	15.0 mw
905 sh	903 sh	919 sh	15.4 mw	15.3 w	
928 vs	918 vs	925 vs		15.9 vs	15.8 m
985 sh	980 vw	978 vw	16.7 w		
				17.4 w	
1014 vs		1016 s	17.8 s	17.8 w	
1026 sh	1027 vs	1034 vs	18.4 s	18.4 m	18.6 ms
		1044 sh			19.0 mw
		1065 m	19.7 ms	19.6 w	20.0 mw
1090 s, b	1093 s, b	1098 m	21.0 ms	21.0 mw	21.0 s
1104 sh	1103 sh	1118 m	21.3 ms	21.3 w	
1154 w	1155 sh	1150 s	22.2 s	22.1 mw	
1170 w	1171 sh	1166 sh			22.5 s
					22.8 w
		1220 s	23.2 m	23.2 w	23.2 w
		1235 w	23.8 ms		
1241 w					
1306 s	1310 s	1321 s	24.6 mw	24.6 vw	
1383 w, sh	1381 w, b	1386 b	25.2 w	25.3 s	
1398 b	1400 b			26.5 ms	26.3 w
1425 b, s	1425 b, s	1420 b, s	27.0 w	27.1 mw	
		1456 sh	27.5 m		
1485 sh	1487 sh	1483 w, sh			27.9 vs
1585 w			28.5 w		
1630 b	1650 b	1630 w, b	29.3 m	29.3 mw	29.6 ms
		2270 w			30.0 mw
2340 b	2342 b	2365 w		30.5 mw	
	2500 w, b	2550 w	31.2 m		
		2798 vs	32.4 w	32.3 vw	32.5 mw
2890 vw			33.2 w		
		2930 vw	35.2 w	35.7 w	35.3 mw
2987 w	2988 sh	2990 w	37.2 w		
3057 sh	3055 sh	3057 sh		37.8 w	
		3070 m			38.3 w
3317 vs	3385 vs	3335 vs	40.1 w		
3420 sh	3455 sh	3450 sh		41.4 mw	
					47.4 mw
			48.2 mw		

Das von Peterson [201)] beschriebene Produkt war also ein Gemisch aus dem Penta- und dem Heptaborat, ähnlich, wie es auch bei der Boresterhydrolyse in Gegenwart von Tetra-n-propylammoniumhydroxid anfällt. Tabelle 6 zeigt einen Vergleich der IR- und Röntgendaten der Benzyldimethyl-n-propylammonium-(1:5)- und (1:7)-Borate und des Benzyldimethylammoniumheptaborats (s. u.). Die Röntgendaten des nicht tabellierten Gemisches enthalten Absorptionslinien, die sowohl dem 1:5- als auch dem 1:7-Borat gemeinsam sind.

3.9.6. Heptaborate tertiärer Amine

Die Boresterhydrolyse in Gegenwart von Tri-n-butylamin als organischer Base führt nach dem Eindampfen im Vakuum zu $[(C_4H_9)_3NH]$ $[B_7O_6(OH)_{10}]$, das auch aus wäßriger Lösung von Base und Borsäure durch Eindampfen im Vakuum und Extrahieren mit siedendem Aceton erhalten wird. Sowohl die IR-Spektren (Tabelle 4) als auch die Röntgendiffraktionsaufnahmen (Tabelle 5) der beiden auf verschiedene Weise dargestellten Salze stimmen recht gut überein. Die Salze sind in Methanol noch einigermaßen gut löslich, jedoch kaum noch in absolutem Äthanol.

Aus wäßriger Lösung von Borsäure und dem tertiären Benzyldimethylamin erhält man durch Eindampfen im Rotationsvakuumverdampfer und Umkristallisieren des Rückstandes aus Wasser oder besser durch Extrahieren des Rückstandes mit trockenem Aceton $[C_6H_5CH_2$ $(CH_3)_2NH][B_7O_6(OH)_{10}]$, dessen IR- und Röntgendaten Tabelle 6 bringt.

Auch bei diesen Polyboraten zwingt offenbar ein großer Raumbedarf des Kations zur Bildung von Heptaboraten.

Mit den stärkeren Basen Tetraalkylammoniumhydroxid, Guanidin ($pKa = 13{,}59$) [205)], Piperidin ($pKa = 11{,}13$) [205)], Tri-n-butylamin ($pKa = 9{,}93$) [205)], Ammoniak ($pKa = 9{,}25$) [206)] und Dimethylbenzylamin ($pKa = 8{,}93$) [205)] werden also definierte Polyborate erhalten. In Gegenwart schwächerer Basen wie Pyridin ($pKa = 5{,}23$) [205)] oder Urotropin ($pKa = 4{,}9$) [207)] fällt bei der Boresterhydrolyse im organischen Lösungsmittel unter den gleichen Bedingungen dagegen nur Borsäure aus. Diese Basen sind also nicht in der Lage, als Kationen in Polyboraten zu wirken.

4. Schema zur Systematik von Boratstrukturen

Boratstrukturen systematisch einzuordnen, wird aufgrund der großen Vielfalt dieser Strukturen immer ein besonderes Problem bleiben. Das erste Schema für Boratstrukturen hat Dale [5)] 1961 angelegt. Die von Tennyson [208)] auf kristallchemischer Grundlage aufgebaute Systematik der Borate ist nicht allgemein und läßt viele Ausnahmen zu.

Tabelle 7. *Systematik der Boratstrukturen*

Inselborat	Kettenborat	Schichtenborat	Raumnetzborat	Struktureinheit Bor-Atome
$[B(OH)_4]^-$ BO_3^{3-} BO_4^{5-}	$\{BO(OH)_2\}_n^{n-}$ *) $\{BO_2\}_n^{n-}$	$\{BO_3(OH)\}_n^{4n-}$	$\{BO_2\}_n^{n-}$	1
$[B_2O(OH)_6]^{2-}$ $B_2O_5^{4-}$		$\{B_2O_3(OH)_2\}_n^{2n-}$ *)		2
$[B_3O_3(OH)_4]^-$ *) $[B_3O_3(OH)_5]^{2-}$ $[B_3O_3(OH)_6]^{3-}$ *) $B_3O_6^{3-}$ *)	$\{B_3O_4(OH)_2\}_n^{n-}$ $\{B_3O_4(OH)_3\}_n^{2n-}$ $\{B_3O_4(OH)_4\}_n^{3n-}$ *)	$\{B_3O_5\}_n^{n-}$ $\{B_3O_5(OH)\}_n^{2n-}$ $\{B_3O_5(OH)_2\}_n^{3n-}$ *)	$\{B_3O_6\}_n^{3n-}$	3
$[B_4O_4(OH)_5]^-$ *) $[B_4O_5(OH)_3]^-$ *) $[B_4O_4(OH)_6]^{2-}$ *) $[B_4O_5(OH)_4]^{2-}$ $[B_4O_4(OH)_7]^{3-}$ *) $[B_4O_5(OH)_5]^{3-}$ *)	$\{B_4O_6(OH)\}_n^{n-}$ *) $\{B_4O_7\}_n^{2n-}$ *) $\{B_4O_6(OH)_2\}_n^{2n-}$ $\{B_4O_6(OH)_3\}_n^{3n-}$ *)	$\{B_4O_7\}_n^{2n-}$		4
$[B_5O_6(OH)_4]^-$ $[B_5O_6(OH)_5]^{2-}$ *) $[B_5O_6(OH)_6]^{3-}$ $[B_5O_6(OH)_7]^{4-}$ *) $B_5O_{10}^{5-}$ *)	$\{B_5O_7(OH)_2\}_n^{n-}$ *) $\{B_5O_7(OH)_3\}_n^{2n-}$ *) $\{B_5O_7(OH)_4\}_n^{3n-}$ $\{B_5O_7(OH)_5\}_n^{4n-}$ *)	$\{B_5O_8\}_n^{n-}$ $\{B_5O_8(OH)\}_n^{2n-}$ $\{B_5O_8(OH)_2\}_n^{3n-}$ $\{B_5O_8(OH)_3\}_n^{4n-}$ *)	$\{B_5O_9\}_n^{3n-}$ *) $\{B_5O_9(OH)\}_n^{4n-}$	5

*) Diese Strukturen wurden noch nicht mit Sicherheit nachgewiesen (vgl. Tabelle 8).

Auch in der hier gegebenen Systematik (Tabelle 7) werden Einschränkungen gemacht, indem nur die Strukturen von Mono-, Di-, Tri-, Tetra- und Pentaboraten eingeordnet werden. Diese Strukturen sind durch Kristallstrukturanalysen, andere Untersuchungsmöglichkeiten wie IR-, Raman- und NMR-Spektren, sowie durch Analogieschlüsse soweit klar, daß sie systematisch zu erfassen sind. Ob alle aufgeführten Strukturen wirklich existieren, hängt von vielen Faktoren ab. Die Systematik ist, das sei klar gesagt, spekulativ aufgestellt, um möglichst vollständig zu sein und noch unbekannte Strukturen einordnen zu können.

Die wenigen bisher bekannten Hexa-, Hepta-, Okta-, Nona- und Dodekaborate, über die aber noch in einem Kapitel dieser Arbeit Bemerkungen gemacht werden, sind in die Systematik noch nicht einbezogen, weil sie, wie es bisher scheint, aus bekannten Struktureinheiten zusammengesetzt sind. Jedoch läßt sich, wie die Anordnung in Tabelle 7 zeigt, eine Einordnung konkreter, nicht zusammengesetzter Anionen bei Vorliegen von mehr Material noch nachholen.

Als *Inselborat-Strukturen* werden solche Strukturen definiert, die nur über Wasserstoffbrücken mit der nächsten Struktureinheit verbunden sind. Die Lage der Wasserstoffatome ist in keiner dieser Strukturen bestimmt, doch muß man sie in Form von OH-Gruppen annehmen. Nun können diese Struktureinheiten durch Austritt von Wasser aber auch zu Gebilden mit O-Brücken kondensieren. Erfolgt der Wasseraustritt nur in einer Richtung, so bilden sich *Ketten*, in zwei Dimensionen *Schichten* und in alle drei Raumrichtungen dreidimensional verknüpfte Gerüste, also *Raumnetzstrukturen*. Eine solche Kondensationsreihe vom Inselpolyborat zum Schichtenpolyborat wurde bei den Calciumtriboraten gefunden:

$$n[B_3O_3(OH)_5]^{2-} \xrightarrow{-nH_2O} \{B_3O_4(OH)_3\}_n^{2n-} \xrightarrow{-nH_2O} \{B_3O_5(OH)\}_n^{2n-}$$

Colemanit-Anion		Meyerhofferit-Anion		synthetisch

Die Nomenklatur dieser Boratstrukturen ist in der Literatur nicht einheitlich. Vorzuschlagen ist, zunächst die Verknüpfungsart (Insel-, Ketten-, Schichtenborat) und dann die Anzahl der Boratome zu nennen, die die Struktureinheit aufbauen. So wäre das Meyerhofferit-Anion ein Kettentriborat, bei dem sich die trimere Einheit laufend wiederholt; also ist Meyerhofferit Calcium-trihydroxo-tetraoxo-triborat-Polymeres.

Diese Art der Nomenklatur führt aber dazu, den Begriff der Mono-(Di-, Tri- usw.)-borat-Polymeren einzuführen.

4.1. Monoborate und Monoborat-Polymere

Monoborate sind also solche Borate, die ein Boratom in der die Struktur aufbauenden Einheit enthalten.

Borate mit 1 Boratom als aufbauender Struktureinheit

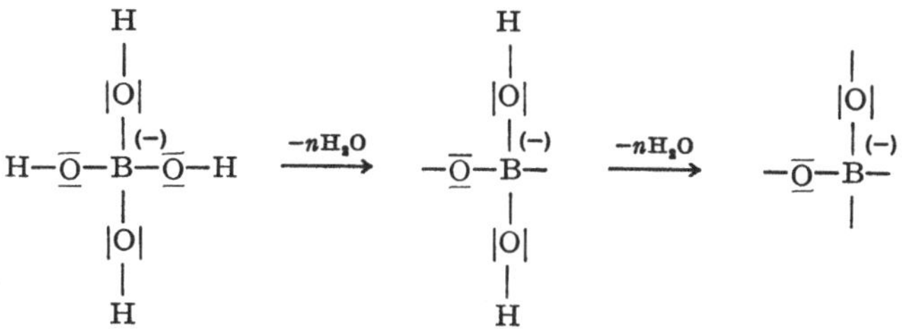

$$\xrightarrow{-nH_2O} \qquad \xrightarrow{-nH_2O}$$

$n[B(OH)_4]^-$ $\{BO(OH)_2\}_n^{n-}$ $\{BO_2\}_n^{n-}$

Inselborat   Kettenborat   Raumnetzborat

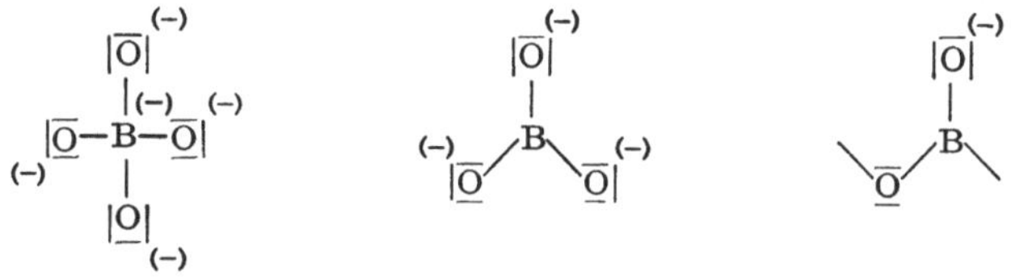

BO_4^{5-} BO_3^{3-} $\{BO_2\}_n^{n-}$

Inselborat   Inselborat   Kettenborat

(aus Schmelzen)   (aus Schmelzen)   (aus Schmelzen)

Borate mit 2 Boratomen als aufbauender Struktureinheit

$$\xrightarrow{-2nH_2O}$$

$n[B_2O(OH)_6]^{2-}$

Inselborat

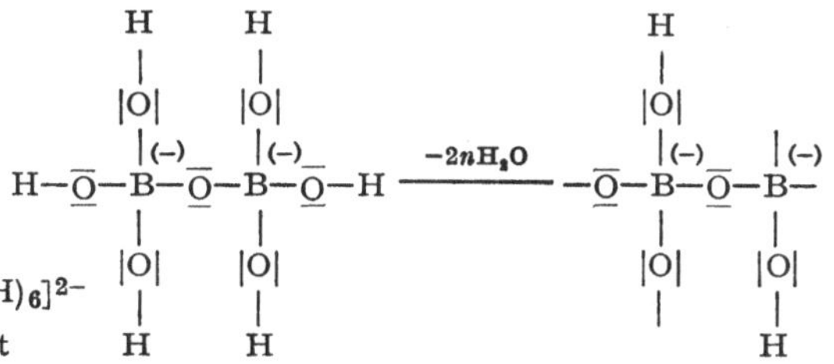

$\{B_2O_3(OH)_2\}_n^{2n-}$

Kettenborat

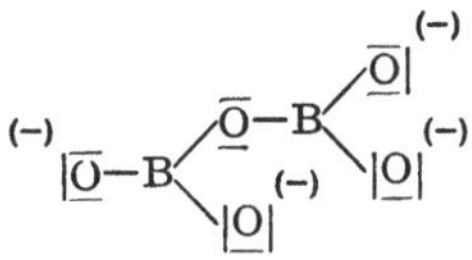

$B_2O_5^{4-}$

Inselborat (aus Schmelzen)

Abb. 6

Das Inselboration $[B(OH)_4]^-$ (Abb. 6) ist in den Monoboraten von Li^+ [61], Na^+ [66], K^+ (wahrscheinlich in $K_2O \cdot B_2O_3 \cdot 8 H_2O$), Rb^+ (analog K^+), Cs^+ (wahrscheinlich in $\alpha\text{-}Cs_2O \cdot B_2O_3 \cdot 8 H_2O$ und in dem durch Esterhydrolyse dargestellten $Cs_2O \cdot B_2O_3 \cdot 4 H_2O$), Ca^{2+} (im rhombischen $CaO \cdot B_2O_3 \cdot 4 H_2O$ und wahrscheinlich in $CaO \cdot B_2O_3 \cdot 6 H_2O$), Sr^{2+} [169] und Ba^{2+} [186] existent.

Das daraus durch intramolekulare Wasserabspaltung abgeleitete Kettenboration $\{BO(OH)_2\}_n^{n-}$ soll in $Cs_2O \cdot B_2O_3 \cdot 2 H_2O$ [113] und in $CaO \cdot B_2O_3 \cdot 2 H_2O$ [144] vorliegen.

Endglied dieser Reihe ist eine Raumnetzstruktur $\{BO_2\}_n^{n-}$, wie sie in $\beta\text{-}LiBO_2$ vorliegt [62].

Ebenfalls mit einem tetraedrisch koordinierten vierbindigem Boratom ist das ungewöhnliche BO_4^{5-}-Ion in der durch Schmelzen bei 1200 °C dargestellten Verbindung $Fe_3^{III}O_2BO_4$ zu formulieren; BO_4^{5-} hat nämlich eine ähnliche Struktur wie $[Al_3Si_3O_{12}]^{3-}$ in dem Silicat Sodalith [196].

In den aus Schmelzen erhaltenen wasserfreien Boraten entspricht die planare BO_3^{3-}-Gruppe, wie sie in Verbindungen mit Aragonit-, Calcit-, Vaterit- oder Dolemitstruktur vorkommt, der CO_3^{2-}-Gruppe in Carbonaten.

Ein wasserhaltiges Borat besitzt die $\{BO_3(OH)\}_n^{4n-}$-Gruppe, nämlich das Mineral Hambergit $\{Be_2(OH)(BO_3)\}_n$ [121] bzw. sein Monohydrat [120]. Schließlich findet man planare Konfiguration um 1 Boratom auch im Kettenion $\{BO_2\}_n^{n-}$, wie es in dem bei 580 °C entstehenden $\alpha\text{-}LiBO_2$ vorkommt [63].

4.2. Diborate und Diborat-Polymere

Das Inseldiboration $[B_2O(OH)_6]^{2-}$ (Abb. 6) existiert im Mineral Pinnoit, $MgO \cdot B_2O_3 \cdot 3 H_2O$ [125], und vermutlich auch in $CaO \cdot B_2O_3 \cdot 6 H_2O$ [148]. Man kann sich dieses Ion aus zwei Molekülen $[B(OH)_4]^-$ durch intermolekularen Austritt von einem Molekül Wasser entstanden denken.

Ein mögliches Schichtendiboration $\{B_2O_3(OH)_2\}_n^{2n-}$ ist nicht bekannt. Das Endprodukt dieser Reihe, das wasserfreie $B_2O_4^{2-}$-Ion, wird für das Calciummetaborat CaB_2O_4 formuliert [167]; es ist seiner Struktur nach identisch mit dem vorher beschriebenen Kettenion $\{BO_2\}_n^{n-}$ und bildet Zickzack-Ketten; die gleiche Struktur wie CaB_2O_4 hat $\gamma\text{-}LiBO_2$ [62].

Ein weiteres wasserfreies Diborat enthält die planare $B_2O_5^{4-}$-Gruppe, die in $Co_2B_2O_5$, $Mg_2B_2O_5$ [137], $Sr_2B_2O_5$ [179], $Fe_2B_2O_5$, $Cd_2B_2O_5$ und ThB_2O_5 enthalten ist.

4.3. Triborate und Triborat-Polymere

Von der kubischen Metaborsäure-I, deren Ringstruktur $B_3O_3(OH)_3$ von Zachariasen [209] aufgeklärt wurde, leiten sich durch sukzessive Aufnahme von OH^--Gruppen das einfach negativ geladene Triboration $[B_3O_3(OH)_4]^-$, das zweifach negativ geladene Triboration $[B_3O_3(OH)_5]^{2-}$ und das dreifach geladene $[B_3O_3(OH)_6]^{3-}$ ab (Abb. 7). Während die Ringstrukturen mit einem und zwei vierbindigen Boratomen auch in Lösung sehr stabil sind, sind die Ringe mit drei vierbindigen Boratomen, entsprechend den Postulaten von Edwards und Ross [4], unter Hydratbedingungen gegenüber der monomeren Form $[B(OH)_4]^-$ instabil.

Kaliummetaborat, $K_2O \cdot B_2O_3 \cdot 2{,}67 H_2O$, das auch als $K_2O \cdot B_2O_3 \cdot 2{,}5 H_2O$ formuliert wird, soll nach Lehmann und Gaube [7] die Struktur $K_3(H_2O)[B_3O_3(OH)_6]$ besitzen, jedoch sprechen unsere Protonenresonanzuntersuchungen [72] eher für ein Schichtenpolyboration der Form $\{[K(H_2O)]_3[B_3O_5(OH)_2]\}_n$. Für das wasserärmere Produkt $K_2O \cdot B_2O_3 \cdot 0{,}67 H_2O$ wird von uns die Strukturformel $\{K_3[B_3O_5(OH)_2]\}_n$ vorgeschlagen. Endglied dieser Reihe, in der das Kettenpolyboration $\{B_3O_4(OH)_4\}_n^{3n-}$ unbekannt ist, ist das $B_3O_6^{3-}$-Ion. Dieses wurde in den wasserfreien Boraten $Na_3B_3O_6$ [88], $K_3B_3O_6$ [98] und im rhomboedrischen $Ba_3(B_3O_6)_2$ [189] durch Kristallstrukturanalyse nachgewiesen.

Das Inselpolyboration $[B_3O_3(OH)_5]^{2-}$ (Abb. 7) kommt in Inderit [127], Kurnakovit [128], α- und β-$Mg(H_2O)_6[B_3O_3(OH)_5]$, in Inderborit [140], $MgCa(H_2O)_6[B_3O_3(OH)_5]_2$, in Inyoit [152], $Ca(H_2O)_4[B_3O_3(OH)_5]$, in synthetischem $Ca(H_2O)_2[B_3O_3(OH)_5]$ [153] und in Meyerhofferit [154], $Ca(H_2O)[B_3O_3(OH)_5]$, vor.

Wahrscheinlich ist auch, daß es in dem durch Esterhydrolyse dargestellten Guanidinium-(2:3)-Borat [202] vorliegt. Das durch Austritt von einem Mol Wasser pro Molekül gebildete Kettenpolyboration $\{B_3O_4(OH)_3\}_n^{2n-}$ (Abb. 7) ist in Hydroboracit [141], $\{MgCa(H_2O)_3[B_3O_4(OH)_3]_2\}_n$ und in Colemanit [156], $\{Ca(H_2O)[B_3O_4(OH)_3]\}_n$, nachgewiesen. Das Schichtenpolyboration $\{B_3O_5(OH)\}_n^{2n-}$ wurde bei der Röntgenstrukturanalyse von synthetischem $\{CaB_3O_5(OH)\}_n$ entdeckt [158]. Das einfach negativ geladene $[B_3O_3(OH)_4]^-$ ist das in konzentrierten Natriumboratlösungen bei p_H 9 vorherrschende Polyion [34]. Es scheint im Mineral Gowerit [161] sowie in den durch Esterhydrolyse dargestellten Ammonium- [42] und Kaliumtriboraten [108] vorzuliegen. Das sich aus diesem Ion ableitende Kettentriboration $\{B_3O_4(OH)_2\}_n^{n-}$ liegt in der monoklinen Metaborsäure-II, $\{H[B_3O_4(OH)_2]\}_n$, nach einer Kristallstrukturanalyse vor [210]. Es leitet sich vom kettenförmigen $\{B_3O_4(OH)_3\}_n^{2n-}$ ab, indem ein vierbindiges Bor unter Verlust einer OH-Gruppe dreibindig wird (Abb. 7).

Für das wasserfreie CsB_3O_5 wurde von Krogh-Moe [58] die Struktur aufgeklärt. Das Ion $\{B_3O_5\}_n^{n-}$ baut ein Schichtenpolyborat auf. Ähnliche Strukturen werden für $Rb_2O \cdot 3 B_2O_3$, $K_2O \cdot 3 B_2O_3$, $Na_2O \cdot 3 B_2O_3$ sowie für das von uns dargestellte $\{NH_4B_3O_5\}_n$ [41] erwartet.

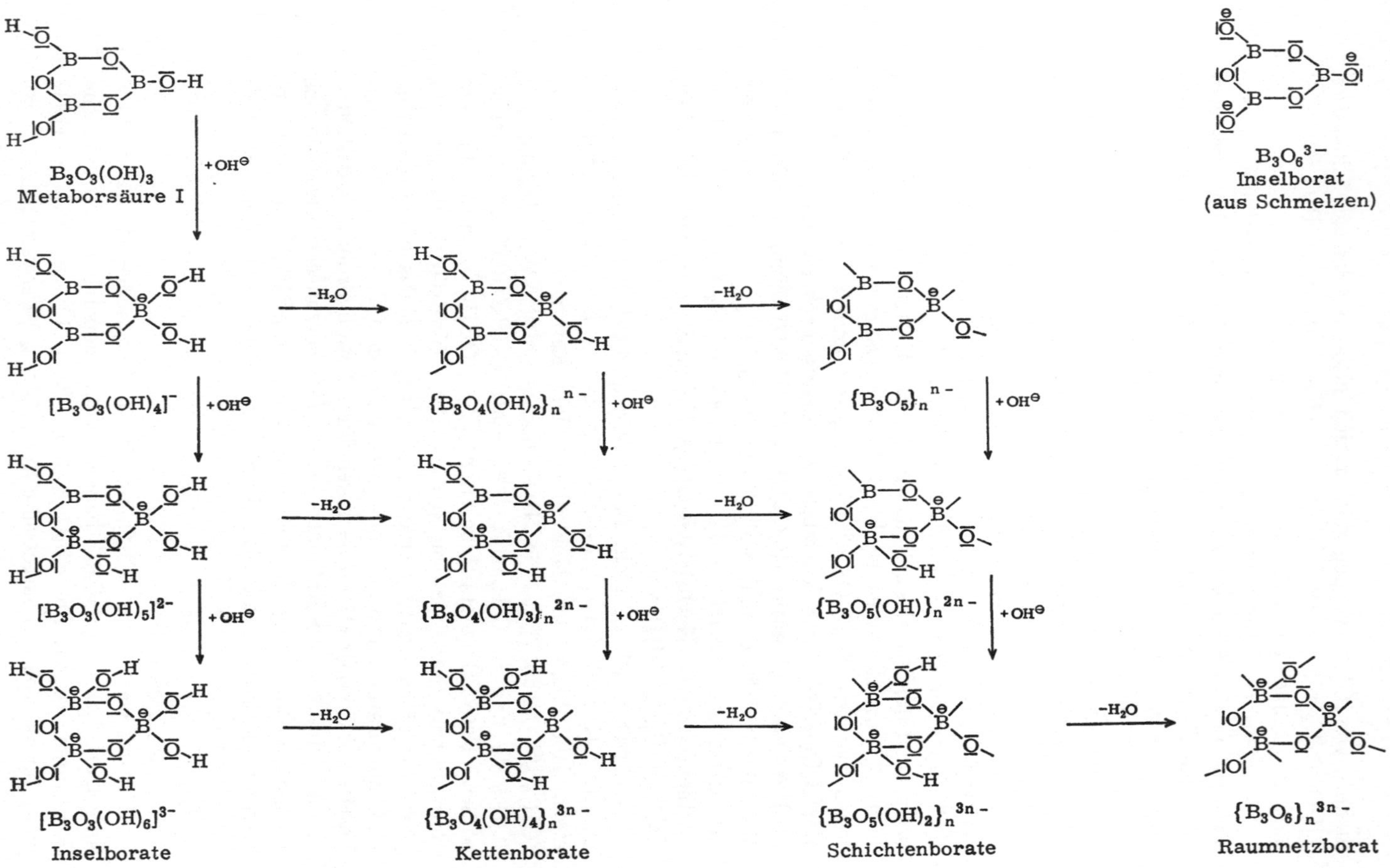

Abb. 7. Borate mit 3 Boratomen als aufbauender Struktureinheit

4.4. Tetraborate und Tetraborat-Polymere

Die denkbaren Tetraboratstrukturen sind in Abb. 8 aufgeführt. Das Inselion $[B_4O_4(OH)_5]^-$ wurde bisher durch Strukturanalyse noch nicht nachgewiesen. Es wird jedoch in folgenden Mineralien oder synthetischen Verbindungen vermutet:

in Paternoit, $MgO \cdot 4\,B_2O_3 \cdot 4\,H_2O$,

in $MeO \cdot 4\,B_2O_3 \cdot 2\,H_2O$ (Me = Ca, Sr, Ba) [113],

in $(NH_4)_2O \cdot 4\,B_2O_3 \cdot 6\,H_2O$ [49],

in $Li_2O \cdot 4\,B_2O_3 \cdot 10\,H_2O$ [5]

sowie in den durch Boresterhydrolyse dargestellten Ammonium- [41] und Piperidinium-(1:4)-Boraten [202].

Theoretisch leiten sich von diesem Ion durch Wasseraustritt das Inselborat $[B_4O_5(OH)_3]^-$ und das Kettenborat $\{B_4O_6(OH)\}_n^{n-}$ ab, die aber beide noch unbekannt sind.

Das Anion $[B_4O_5(OH)_4]^{2-}$ ist das wichtigste und *bekannteste Polyboration*, denn es wurde zuerst im häufigsten Bormineral, nämlich in *Borax*, durch Kristallstrukturanalyse [71] nachgewiesen. Außerdem wurde diese Struktureinheit bei Strukturuntersuchungen im *Tinkalkonit* [73] und im synthetischen $K_2O \cdot 2\,B_2O_3 \cdot 4\,H_2O$ [55] gefunden sowie für das Mineral *Halurgit*, $MgO \cdot 2\,B_2O_3 \cdot 5\,H_2O$ [131], für die aus wäßrigen Lösungen synthetisierten Verbindungen $(NH_4)_2O \cdot 2\,B_2O_3 \cdot 4\,H_2O$ [48], $Li_2O \cdot 2\,B_2O_3 \cdot 4\,H_2O$ und $Li_2O \cdot 2\,B_2O_3 \cdot 3\,H_2O$ sowie ferner für das durch Esterhydrolyse dargestellte Kalium-(1:2)-Borat [108] vorausgesagt. Ein wasserreicheres $[B_4O_4(OH)_6]^{2-}$ ohne O—B—O-Brücke, das sich aus $[B_4O_4(OH)_5]^-$ durch Anlagerung einer OH^--Gruppe bilden kann, wird der Verbindung $MgO \cdot 2\,B_2O_3 \cdot 9\,H_2O$ zugeschrieben [135]. Dagegen ist das wasserärmere Anion $\{B_4O_6(OH)_2\}_n^{2n-}$, das eine Kettenstruktur besitzen muß, für das Mineral *Kernit* vorausgesagt und nachgewiesen [75,79] worden. Auch das durch thermischen Abbau von $K_2O \cdot 2\,B_2O_3 \cdot 2\,H_2O$ gewonnene $K_2O \cdot 2\,B_2O_3 \cdot H_2O$ enthält dieses Ion, wie NMR-Untersuchungen [72] gezeigt haben.

Endglied dieser Reihe ist das Ion $\{B_4O_7\}_n^{2n-}$, das zum Beispiel in $Li_2O \cdot 2\,B_2O_3$ [64] oder in $CdO \cdot 2\,B_2O_3$ [211] Schichten aufbaut. Nach ^{11}B—NMR-Untersuchungen an Abbauprodukten von Kernit muß außerdem eine Kettenform des Ions $\{B_4O_7\}_n^{2n-}$ möglich sein.

Das Anion $[B_4O_5(OH)_5]^{3-}$ scheint im Guanidinium-(3:4)-Borat vorzuliegen, das durch Boresterhydrolyse [202] gewonnen wurde. Die sich daraus ableitenden Insel-, Ketten- und Schichtenstrukturen $[B_4O_4(OH)_7]^{3-}$, $\{B_4O_6(OH)_3\}_n^{3n-}$ und $\{B_4O_7(OH)\}_n^{3n-}$ sind bisher noch unbekannt.

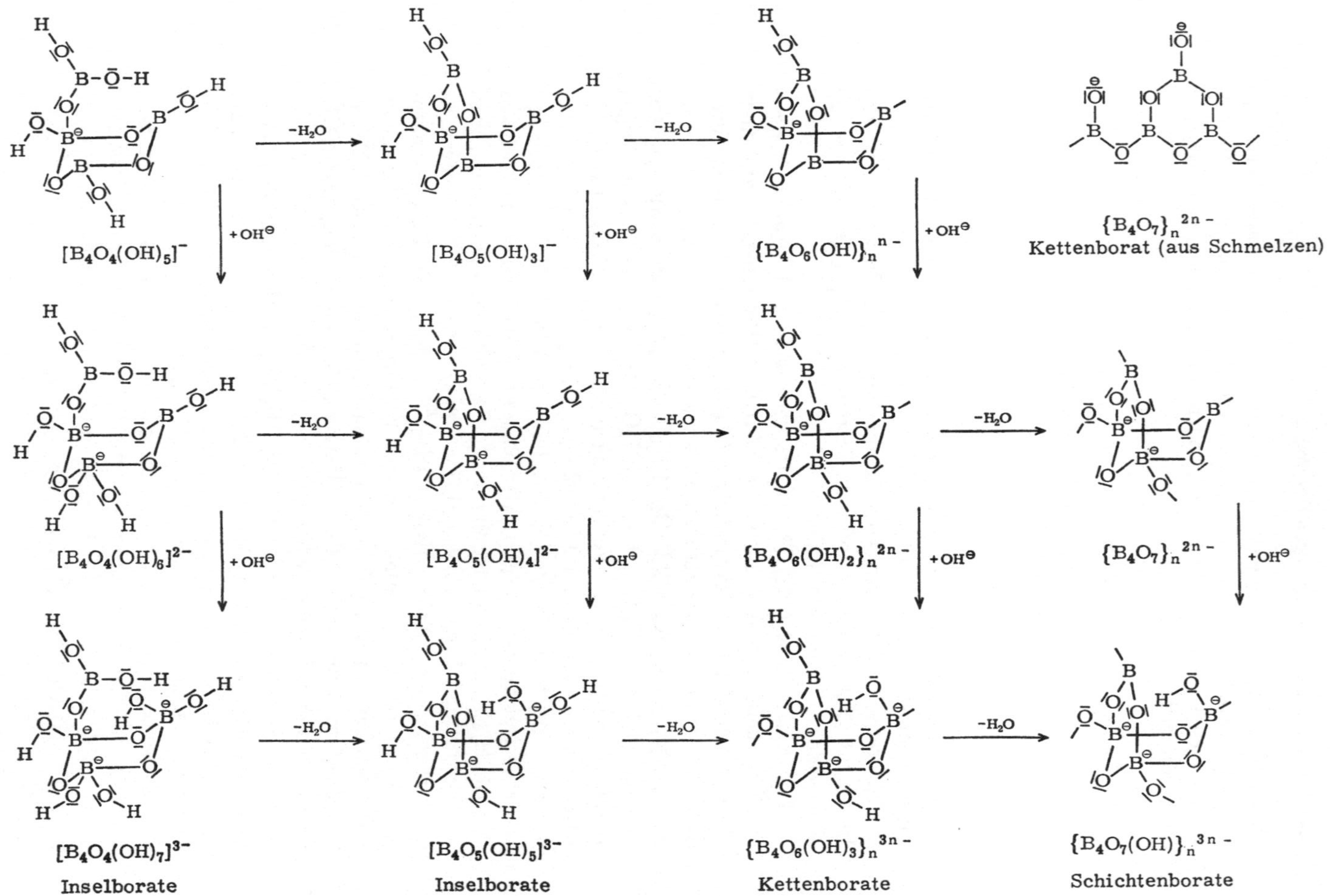

Abb. 8. Borate mit 4 Boratomen als aufbauender Struktureinheit

4.5. Pentaborate

Die denkbaren Pentaboratstrukturen zeigt Abb. 9. Die Existenz des Ions $[B_5O_6(OH)_4]^-$ wurde im synthetischen $K_2O \cdot 5\,B_2O_3 \cdot 8\,H_2O$ auf vielfältige Art und Weise [52,53,54,103,104] sichergestellt. Wahrscheinlich liegt dieses Ion auch im Mineral

$$\text{Sborgit, } Na_2O \cdot 5\,B_2O_3 \cdot 10\,H_2O,$$

$$\text{in } Li_2O \cdot 5\,B_2O_3 \cdot 10\,H_2O,$$

$$\text{in } Rb_2O \cdot 5\,B_2O_3 \cdot 8\,H_2O \text{ und}$$

$$\text{in } Cs_2O \cdot 5\,B_2O_3 \cdot 8\,H_2O \text{ sowie}$$

in den durch Boresterhydrolyse dargestellten (1:5)-Boraten von Kalium [108], Ammonium [41], Guanidinium, Piperidinium, Tetramethylammonium, Tetraäthylammonium und Tetrapropylammonium [202] vor. Beim thermischen Abbau von $K(H_2O)_2[B_5O_6(OH)_4]$ entsteht, wie NMR-Spektren [72] zeigen, in $K_2O \cdot 5\,B_2O_3 \cdot 2\,H_2O$ das Kettenpolyion $\{B_5O_7(OH)_2\}_n^{n-}$. Endglied dieser Reihe ist das Ion $\{B_5O_8\}_n^{n-}$, das in der Kaliumverbindung Helixketten aufbaut, die mit Nachbarketten Schichten bilden [106,107]; ähnlich soll die Struktur von $\{RbB_5O_8\}_n$ sein.

Das Ion $[B_5O_6(OH)_5]^{2-}$ kann im Mineral *Ezcurrit*, $2\,Na_2O \cdot 5\,B_2O_3 \cdot 7\,H_2O$ [82], in Augers Borat, $2\,K_2O \cdot 5\,B_2O_3 \cdot 5\,H_2O$ [66], und in dem durch Esterhydrolyse gewonnenen Kalium-(2:5)-Borat [108], $K_2[B_5O_6(OH)_5]$, enthalten sein. Das sich daraus ableitende Kettenpentaboration $\{B_5O_7(OH)_3\}_n^{2n-}$ soll im Mineral *Nasinit*, $2\,Na_2O \cdot 5\,B_2O_3 \cdot 5\,H_2O$, vorkommen. Das Schichtenpentaboration $\{B_5O_8(OH)\}_n^{2n-}$ ist in *Veatchit* und *Paraveatchit* enthalten [175,176].

Das Vorliegen des Ions $[B_5O_6(OH)_6]^{3-}$ im Mineral *Ulexit* wurde durch Röntgenstrukturanalyse [92] nachgewiesen. Ebenso eindeutig ist, daß das Kettenpolyion $\{B_5O_7(OH)_4\}_n^{3n-}$ im Mineral *Probertit* [91] strukturbestimmend ist. Dieses Ion wird auch für das Mineral *Preobrazhenskit*, $3\,MgO \cdot 5\,B_2O_3 \cdot 4^{1}/_2\,H_2O$, vorausgesagt. Das Schichtenion dieser Reihe, $\{B_5O_8(OH)_2\}_n^{3n-}$, liegt im Mineral *Heidornit*, $\{Ca_3Na_2Cl(SO_4)_2[B_5O_8(OH)_2]\}_n$ [212], vor; die Raumnetzstruktur $\{B_5O_9\}_n^{3n-}$ ist noch unbekannt.

Das Inselpolyboration $[B_5O_6(OH)_7]^{4-}$ könnte in den Mineralien *Pandermit*, $4\,CaO \cdot 5\,B_2O_3 \cdot 7\,H_2O$, und *Tertschit*, $4\,CaO \cdot 5\,B_2O_3 \cdot 20\,H_2O$, sowie in dem durch Boresterhydrolyse gewonnenen Guanidinium-(4:5)-Borat enthalten sein. Während ein Kettenpentaborat $\{B_5O_7(OH)_5\}_n^{4n-}$ noch unbekannt ist, ist das Schichtenpentaboration $\{B_5O_8(OH)_3\}_n^{4n-}$, allerdings bei Ersatz einer OH^--Gruppe durch ein Cl^-, in den Mineralien der Hilgarditgruppe zu finden; so wird für *Hilgardit* [213] die Strukturformel $\{Ca_2[B_5O_8(OH)_2]Cl\}_n$ gefordert. Die Raumnetzstruktur $\{B_5O_9(OH)\}_n^{4n-}$ wird für das Mineral *Howlit*, $\{Ca_2[B_5O_9(OH)] \cdot Si(OH)_4\}_n$, postuliert [214].

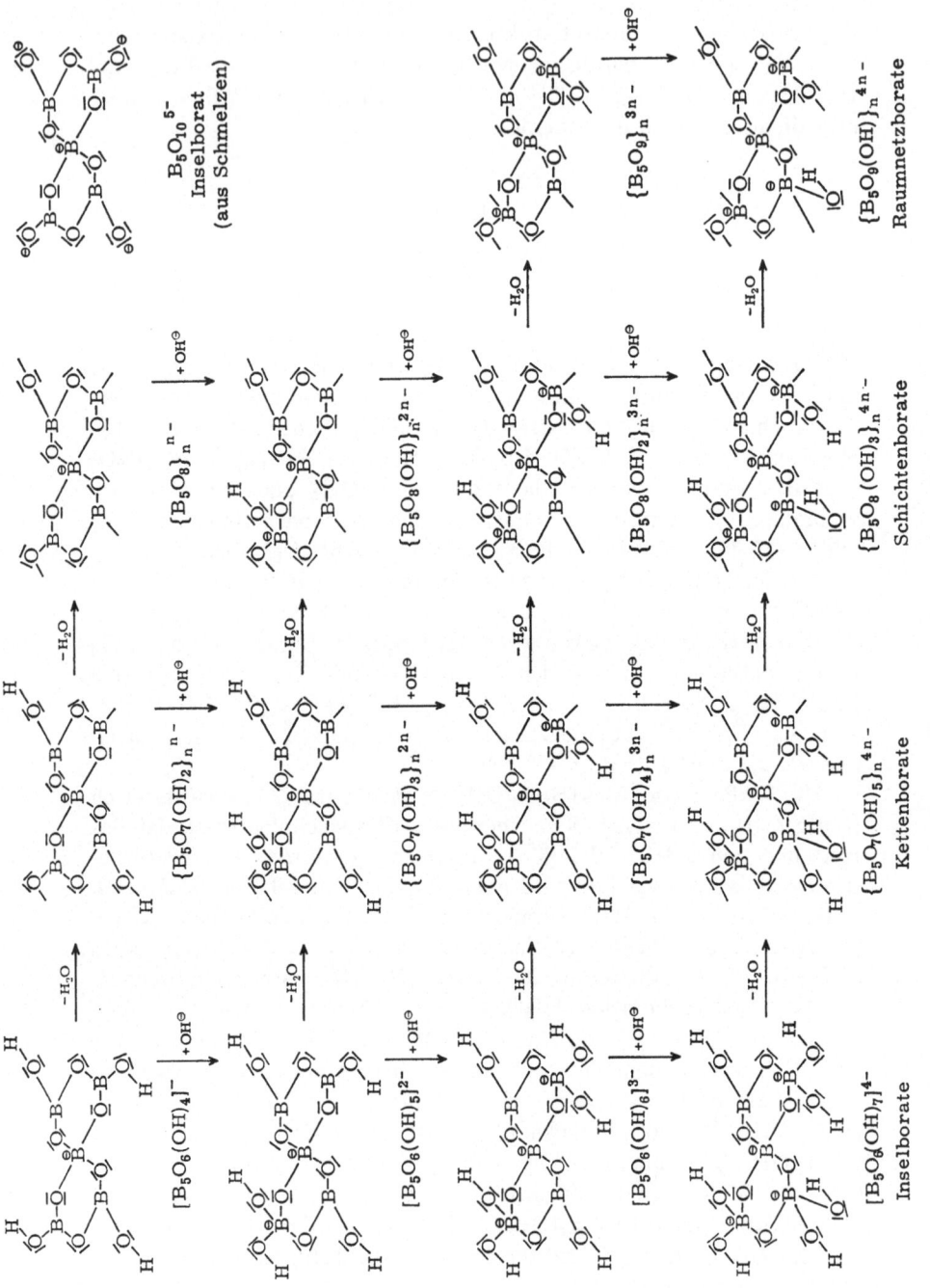

Abb. 9. Borate mit 5 Boratomen als aufbauender Struktureinheit

4.6. Höhere Borate

Im *Ginorit* soll das Inselhexaboration $[B_6O_7(OH)_6]^{2-}$ vorliegen; wenn dieses Prognose bestätigt werden kann, wäre es aus dem bekannten Inseltriboration $[B_3O_3(OH)_4]^-$ durch Abspaltung eines H_2O-Moleküls aus zwei Molekülen des Anions entstanden und könnte als Doppeltriborat gelten.

Im *Kaliborit* [110)] liegen Anioneneinheiten der Form $\{B_6O_8(OH)_5\}_n^{3n-}$ vor, die sich aus einer B_5-Struktureinheit und einer $BO(OH)_2$-Gruppe zusammensetzen.

Ein weiteres, bereits durch Röntgenstrukturanalyse [8)] und NMR-Spektroskopie [9)] im *Tunellit*, $SrO \cdot 3 B_2O_3 \cdot 4 H_2O$, aufgeklärtes Hexaboration ist das Schichtenboration $\{B_6O_9(OH)_2\}_n^{2n-}$ mit einem dreibindigen $O^{(+)}$-Atom in der Mitte dreier Sechsringsysteme. Dieses Sauerstoffatom ist mit drei vierbindigen Boratomen verbunden. Aufgrund der Elektronegativitätsunterschiede von O und B ist aber denkbar, daß eine der lockeren Bindungen zwischen $O^{(+)}$ und $B^{(-)}$ aufklappt und eines der drei Wassermoleküle, die durch Wasserstoffbrückenbindungen die einzelnen Schichten zusammenhalten, stärker an eines der Boratome herangezogen wird. Berücksichtigt man diese sehr wahrscheinliche Möglichkeit des Aufbaus der Struktur des Tunellit-Anions, so kann man diese Struktur auf das Kettentriboration $\{B_3O_4(OH)_2\}_n^{n-}$ zurückführen und müßte nicht von einem „einzigartigen Anion" in der Boratchemie [8)] sprechen. Dieses Anion soll ja auch im isostrukturellen Mineral Nobleit, $CaO \cdot 3 B_2O_3 \cdot 4 H_2O$, vorliegen [160)]. Im wasserfreien $SrO \cdot 2 B_2O_3$ [180)] ist ebenfalls ein dreibindiges Sauerstoffatom vorhanden; hier sind alle Boratome vierbindig. Es bildet eine Schichtenstruktur aus Ketten sechsgliedriger Ringe mit gemeinsamen B—O-Kanten aus.

Ein weiterer Strukturtyp findet sich im synthetischen Metaborat $Zn_4O[B_6O_{12}]$ [196)] und in der isostrukturellen Hg-Verbindung [197)]. Die Struktur dieser beiden Verbindungen besteht aus einer Raumnetzstruktur, die aus großen $B_6O_{12}^{6-}$-Ringen gebildet wird. Alle Boratome befinden sich in Tetraedern, die Kanten und Ecken gemeinsam haben; diese Struktur ist dem dreidimensionalen Gerüst des $(Al_3Si_3O_{12})^{3-}$-Ions im Mineral Sodalith sehr ähnlich. Auf diese Weise werden „*Käfige*" ähnlich dem Ultramarin-Gerüst aus vier Tetraederringen $[B_6O_{18}]$ und sechs Tetraederringen $[B_4O_{12}]$ aufgebaut; das zusätzliche Sauerstoffatom befindet sich im Zentrum dieses Käfigs.

Eine Reihe von Heptaboraten mit den organischen Kationen Tetra-n-propylammonium-, Tetra-n-butylammonium-, Dimethylbenzyl-n-propylammonium, Tri-n-butylammonium und Dimethylbenzylammonium [202)] wurde durch Boresterhydrolyse erhalten. Die Struktur des Inselpolyborations $[B_7O_6(OH)_{10}]^-$ ist schwer vorstellbar, und es ist gut möglich, daß sich dieses Ion aus zwei schon bekannten Inselpolyborationen zusammensetzt, die über Wasserstoffbrücken miteinander verbunden sind. Ein weiteres Heptaborat liegt im Mineral *Boracit*, $Mg_3B_7O_{13}Cl$ [139] vor, das jedoch noch ein Cl-Atom pro Formeleinheit enthält. Das Oktaboration $B_8O_{14}^{4-}$, das im wasserfreien $BaO \cdot 2 B_2O_3$ vorliegen soll, ist in Wirklichkeit, wie die Kristallstrukturanalyse [190] zeigt, aus einem Pentaboration und einem Triboration zusammengesetzt. Diese Strukturfolge wiederholt sich und bildet eine Raumnetzstruktur aus.

Dunicz [215] schlägt für die wasserfreien Polyborate $B_8O_{13}^{2-}$, $B_6O_{10}^{2-}$ und $B_4O_7^{2-}$ jeweils Käfigstrukturen vor, bei denen ein O^{2-}-Ion eingeschlossen ist.

Ein zusammengesetzter Strukturtyp wird auch für das durch Esterhydrolyse dargestellte $K[B_{12}O_{15}(OH)_7]$ oder $K[B_{12}O_{16}(OH)_5]$ erwartet. Bei den höheren Boraten finden wir also bisher keinen Anhaltspunkt dafür, daß sie nicht mindestens durch intramolekularen Wasseraustritt aus den bekannten Tri-, Tetra- oder Pentaboratstrukturtypen abzuleiten wären.

Daher wurde versucht, in der Systematik nur alle denkbaren Mono-, Di-, Tri-, Tetra- und Pentaborate sowie ihre Polymere zu berücksichtigen.

Tabelle 8 (siehe S. 275) zeigt, welche Boratstrukturen bereits als gesichert anzusehen sind, und wieviele Borate strukturchemisch noch nicht aufgeklärt sind, welche Arbeit also noch zu leisten bleibt.

Tabelle 8. *Systematik bekannter Boratstrukturen*

Inselborat	Kettenborat	Schichtenborat	Raumnetzborat	Aufbauende Bor-Atome
$Na(H_2O)_4[B(OH)_4]$ Natriummonoborat 68)	$\beta-\{Cs(H_2O)_3[BO(OH)_2]\}_n$? synthetisch 113)	$\{Be_2[BO_3(OH)] \cdot H_2O\}_n$ 120)	$\beta-\{LiO_2\}_n$ 62)	1
$LaBO_3$ 192)	$\{Ca(BO_2)_2\}_n$ 167)			
$Fe_3O_2(BO_4)$ 195)				
$Mg[B_2O(OH)_6]$ Pinnoit 125)				2
$Mg_2B_2O_5$ 137)				
$Ca(H_2O)[B_3O_3(OH)_4]_2$? Gowerit 161)	$\{H[B_3O_4(OH)_2]\}_n$ Metaborsäure II 209)	$\{CsB_3O_5\}_n$ 58)	$\{Na_3B_3O_6\}_n$ 88)	3
$Ca(H_2O)[B_3O_3(OH)_5]$ Meyerhofferit 154)	$\{Ca(H_2O)[B_3O_4(OH)_3]\}_n$ Colemanit 156)	$\{Ca[B_3O_5(OH)]\}_n$ synthetisch 158)		
		$\{[K(H_2O)]_3[B_3O_5(OH)_2]\}_n$? synthetisch 72)		
$Mg(H_2O)_6[B_4O_4(OH)_6]$? synthetisch 135)	$\{[Na_2(H_2O)_3][B_4O_6(OH)_2]\}_n$ Kernit 79)	$\{Li_2B_4O_7\}_n$ 64)		4
$[Na(H_2O)_4]_2[B_4O_5(OH)_4]$ Borax 71)				
$K(H_2O)_2[B_5O_6(OH)_4]$ synthetisch 52)	$\{[Na_2(H_2O)][B_5O_7(OH)_3]\}_n$? Nasinit 66)	$\{KB_5O_8\}_n$ 109)	$\{Ca_2[B_5O_9(OH)] \cdot Si(OH)_4\}_n$ Howlit 214)	5
$[Na_2(H_2O)][B_5O_6(OH)_5]$? Ezcurrit 82)	$\{[NaCa(H_2O)_3][B_5O_7(OH)_4]\}_n$ Probertit 91)	$\{Sr_2(H_2O)[B_5O_8(OH)]\}_2 \cdot$ $B(OH)_3\}_n$ Veatchit 176)		
$[NaCa(H_2O)_5][B_5O_6(OH)_6]$ Ulexit 92)		$\{Na_2Ca_3Cl(SO_4)_2[B_5O_8(OH)_2]\}_n$ Heidornit 212)		
$[Ca(H_2O)_3]_2[B_5O_6(OH)_7]$? Tertschit 162)				

5. Literatur

1) Edwards, J. O., Morrison, G. C., Ross, V. F., Schultz, J. W.: J. Am. Chem. Soc. 77, 266 (1955).
2) Schorsch, G., Ingri, N.: Acta Chem. Scand. 21, 2727 (1967).
3) Lourijsen-Teyssedre, M.: Bull. Soc. Chim. France 1955, 1111.
4) Edwards, J. O., Ross, V. F.: J. Inorg. Nucl. Chem. 15, 329 (1960).
5) Dale, J.: J. Chem. Soc. (London) 1961, 922.
6) Christ, C. L.: Am. Mineralogist 45, 334 (1960); Proc. VIIth Internat. Conference Coordination Chem., Stockholm and Uppsala, Abstracts, 1962, 178.
7) Lehmann, H.-A., Gaube, W.: Z. Anorg. Allg. Chem. 329, 190 (1964).
8) Clark, J. R.: Science (Washington) 141, 1178 (1963).
9) Cuthbert, J. D., MacFarlan, W. T., Petch, H. E.: J. Chem. Phys. 43, 173 (1965).
10) Silver, A. H., Bray, P. J.: J. Chem. Phys. 29, 948 (1958).
11) Weir, C. E.: J. Res. Nat. Bur. Std. 70A, 153 (1966).
12) Nies, P. N., Hulbert, R. W.: J. Chem. Eng. Data 12, 303 (1967).
13) Carpéni, G., Haladjian, J., Pilard Mlle.: Bull. Soc. Chim. France 1960, 1634.
14) Tolédano, P., Awka, S.: Bull. Soc. Chim. France 1966, 1933.
15) Menzel, H.: Z. Anorg. Allg. Chem. 164, 122 (1927).
16) Mueller, P., Abegg, R.: Z. Physik. Chem. 57, 513 (1907).
17) Souchay, P.: Bull. Soc. Chim. France 1951, 932.
18) Marres, M.: Rev. Chim. Min. 4, 803 (1967).
19) Spiryagina, A. I.: Ber. Akad. Wiss. UdSSR 68, 909 (1949).
20) Schwarz, Je. M., Jevins, A. F.: Nachr. Akad. Wiss. LettSSR 1955, Nr. 7, 127.
21) Everest, D. A., Popiel, W. J.: J. Chem. Soc. (London) 1959, 657.
22) Doucet, Y., Rollin, M.: C. R. Hebd. Séances Acad. Sci. 226, 1967 (1948).
23) Thygesen, J. E.: Z. Anorg. Allg. Chem. 237, 101 (1938).
24) Keschan, A. D., Wimba, Ss. G., Schwarz, Je. M.: Nachr. Akad. Wiss. LettSSR 1956, Nr. 9, 135.
25) Stetten jr., D.: Anal. Chem. 23, 1177 (1951).
26) Edwards, J. O.: J. Am. Chem. Soc. 75, 6154 (1953).
27) Antikainen, P. J.: Suomen Kemistilehti B 30, 74 (1957).
28) Goulden, J. D. S.: Spectrochim. Acta (London) 1959, 657.
29) Hibben, J. W.: Am. J. Sci. 35A, 113 (1938).
30) Onak, T. P., Landesman, H., Williams, R. E., Shapiro, I.: J. Phys. Chem. 63, 1533 (1959).
31) Ingri, N., Lagerström, G., Frydman, M., Sillen, L. G.: Acta Chem. Scand. 11, 1034 (1957).
32) — Proc. VIIth Internat. Conference Coordination Chem., Stockholm and Uppsala, Abstracts, 1962, 182.
33) Carpéni, G., Souchay, P.: J. Chim. Phys. 42, 149 (1945).
34) Ingri, N.: Svensk. Kem. Tidskr. 75, 199 (1963).
35) Anderson, J. L., Eyring, E. M., Whittaker, M. P.: J. Phys. Chem. 68, 1128 (1964).
36) Lefebvre, J.: J. Chim. physique 54, 567 (1957); C. R. Hebd. Séances Acad. Sci. 241, 1295 (1955).
37) Carpéni, G.: Bull. Soc. Chim. France 1955, 1327.
38) Jahr, K. F., Fuchs, J.: Chem. Ber. 96, 2457 (1963).
39) Fuchs, J., Jahr, K. F.: Chem. Ber. 96, 2460 (1963).
40) — — Heller, G.: Chem. Ber. 96, 2472 (1963).
41) Heller, G.: J. Inorg. Nucl. Chem. 27, 2346 (1965).
42) — Z. Anal. Chem. 214, 23 (1965).

43) Clark, J. R., Christ, C. L.: Am. Mineralogist *44*, 1150 (1959).
44) — Am. Mineralogist *45*, 1087 (1960).
45) Atterberg, A.: Z. Anorg. Chem. *48*, 367 (1906).
46) Tolédano, P., Matringe, M. A.: C. R. Hebd. Séances Acad. Sci. *264*, 305 (1967).
47) — C. R. Hebd. Séances Acad. Sci. *266*, 1155 (1968).
48) Sborgi, U., Ferri, L.: Mem. Reale Accad. naz. Lincei, Cl. Sci. fisiche mat. natur [5], *13*, 569 (1922).
49) Menzel, H.: Z. Anorg. Allg. Chem. *166*, 63 (1927).
50) Lehmann, H.-A., Schmidt, W.: Z. Chem. *5*, 111 (1965).
51) — — Z. Chem. *5*, 65 (1965).
52) Zachariasen, W. H., Plettinger, H. A.: Acta Cryst. (Copenhagen) *16*, 376 (1963).
53) Smith, J. A. S., Richards, R. E.: Trans. Faraday Soc. *48*, 307 (1952).
54) Silvidi, A. A., McGrath, J. W.: J. Chem. Phys. *30*, 1028 (1959).
55) Marezio, M., Plettinger, H. A., Zachariasen, W. H.: Acta Cryst. (Copenhagen) *16*, 975 (1963).
56) Filsinger, F.: Arch. Phar., *208*, 211 (1876).
57) Ingri, N.: Acta Chem. Scand. *17*, 581 (1963).
58) Krogh-Moe, J.: Acta Cryst. (Copenhagen) *13*, 889 (1960).
59) Reburn, W. T., Gale, W. A.: J. Phys. Chem. *59*, 19 (1955).
60) Bouaziz, R.: Bull. Soc. Chim. France *1962*, 1451.
61) Höhne, E.: Z. Anorg. Allg. Chem. *342*, 188 (1966).
62) Lehmann, H.-A., Tiess, D.: Chem. Techn. *11*, 260 (1959).
63) Höhne, E., Kutschabsky, L.: Z. Chem. *3*, 33 (1963).
64) Krogh-Moe, J.: Acta Cryst. (Copenhagen) *15*, 190 (1962).
65) Bues, W., Förster, G., Schmitt, R.: Z. Anorg. Allg. Chem. *344*, 148 (1966).
66) Auger, V.: C. R. Hebd. Séances Acad. Sci. *180*, 1602 (1925).
67) Menzel, H., Schulz, H.: Z. Anorg. Allg. Chem. *251*, 167 (1943).
68) Block, S., Perloff, A.: Acta Cryst. (Copenhagen) *16*, 1233 (1963).
69) Krc jr., J.: Anal. Chem. *23*, 806 (1951).
70) Schwarz, Je. M., Grundstein, V. V., Jevins, A. F.: Russ. J. Inorg. Chem. *12*, 1061 (1967).
71) Morimoto, N.: Mineral. J. (Sapporo) *2*, 1 (1956).
72) Jahr, K. F., Wegener, K., Heller, G., Worm, K.: J. Inorg. Nucl. Chem. *30*, 1677 (1968).
73) Christ, C. L., Garrels, R. M.: Am. J. Sci. *257*, 516 (1959).
74) Cuthbert, J. D., Petch, H. E.: J. Chem. Phys. *38*, 1912 (1963).
75) Petch, H. E., Pennington, K. S., Cuthbert, J. D.: Am. Mineralogist *47*, 401 (1962).
76) Dharmatti, S. S., Iyer, S. A., Vijayaraghavan, R.: J. Phys. Soc. Japan *17*, 1736 (1962).
77) Menzel, H., Schulz, H.: Z. Anorg. Allg. Chem. *245*, 157 (1940).
78) Ross, V., Edwards, J. O.: Acta cryst. (Copenhagen) *12*, 258 (1959).
79) Giese jr., R. F.: Science (Washington) *154*, 1453 (1966).
80) Cialdi, G., Corazza, E., Sabelli, C.: Atti Accad. naz. Lincei, Rend., Cl. Sci. fisiche, mat. natur [8] *42*, 236 (1967).
81) Ball, G. R., Kemp, P. H.: Brit. Pat. 737925 (5. Oktober 1955); über C. A. *50*, 8979 (1956).
82) Muessig, S., Allen, R. D.: Econ. Geol. *52*, 426 (1957).
83) Cipriani, C.: Atti Accad. naz. Lincei, Rend., Cl. Sci. fisiche, mat. natur [8] *22*, 519 (1957).
84) Abdullajew, G. K., Mamedow, Ch. S.: Ber. Akad. Wiss. AserbaidschanSSR *22*, 21 (1966).

85) Rollet, A. P., Peng Chung: Bull. Soc. Chim. France [5], *2*, 982 (1935).

86) Ponomareff, J.: Z. Anorg. Allg. Chem. *89*, 383 (1914).

87) Morey, G. W., Merwin, H. E.: J. Am. Soc. *58*, 2248 (1936).

88) Marezio, M., Plettinger, H. A., Zachariasen, W. H.: Acta Cryst. (Copenhagen) *16*, 594 (1963).

89) Hyman, A., Perloff, A., Mauer, F., Block, S.: Acta Cryst. (Copenhagen) *22*, 815 (1967).

90) Clark, J. R., Christ, C. L.: Am. Mineralogist *44*, 712 (1959).

91) Kurbanow, Ch. M., Rumanowa, J. M., Below, N. W.: Ber. Akad. Wiss. UdSSR *152*, 1000 (1963).

92) Clark, J. R., Appleman, D. E.: Science (Washington) *145*, 1295 (1964).

93) Lehmann, H.-A., Gaube, W.: Z. Chem. *1*, 93 (1961).

94) Tolédano, P.: Rev. Chim. Min. *1*, 353 (1964).

95) Dukelski, M.: Z. Anorg. Chem. *50*, 38 (1906).

96) Rollet, A. P., Tolédano, P.: C. R. Hebd. Séances Acad. Sci. *255*, 2261 (1962).

97) Zviedre, I. I., Osol, Ja. K., Jevins, A. F.: Russ. J. Inorg. Chem. *13*, 2613 (1968).

98) Zachariasen, W. H.: J. Chem. Phys. *5*, 919 (1937).

99) Bray, P. J., Edwards, J. O., O'Keefe, J. G., Ross, V., Tatsuzaki, I.: J. Chem. Phys. *35*, 435 (1961).

100) Ssauka, Ja Ja.: Kristallographie (UdSSR) *3*, 93 (1958).

101) Kume, K., Kakiuchi, Y.: J. Phys. Soc. Japan *15*, 329 (1960).

102) Haladjian, J., Carpeni, G.: Bull. Soc. Chim. France *1960*, 1629.

103) Zachariasen, W. H.: Z. Krist., Min., Petrogr., Abt. A *98*, 266 (1937).

104) Lal, K. C., Petch, H. E.: J. Chem. Phys. *43*, 178 (1965).

105) Rosenheim, A., Leyser, F.: Z. Anorg. Allg. Chem. *119*, 1 (1921).

106) Krogh-Moe, J.: Ark. Kemi *14*, 567 (1959).

107) — Acta Cryst. (Copenhagen) *18*, 1088 (1965).

108) Heller, G.: J. Inorg. Nucl. Chem. *29*, 2181 (1967).

109) Wegener, K.: J. Inorg. Nucl. Chem. *29*, 1847 (1967).

110) Corazza, E., Sabelli, C.: Atti Accad. naz. Lincei, Rend., Cl. Sci. fisiche, mat. natur [8], *38*, 707 (1965); [8], *41*, 527 (1966).

111) Rollet, A. P., Andres, L.: Bull. Soc. Chim. France [4] *49*, 847 (1931).

112) Ferrari, A., Magnani, A.: Gazz. Chim. Ital. *69*, 275 (1939).

113) Lehmann, H.-A.: Z. Chem. *3*, 284 (1963).

114) — Gaube, W.: Z. Anorg. Allg. Chem. *335*, 50 (1965).

115) Kocher, J.: Rev. Chim. Min. *3*, 209 (1966); Bull. Soc. Chim. France *1968*, 919.

116) Reischle, A.: Z. Anorg. Chem. *4*, 166 (1893).

117) Tolédano, P.: Bull. Soc. Chim. France *1966*, 2302.

118) Krogh-Moe, J., Ihara, M.: Acta Cryst. (Copenhagen) *23*, 427 (1968).

119) Lehmann, H.-A., Schubert, E.: Z. Chem. *8*, 116 (1968).

120) Schlatti, M.: Naturwissenschaften *54*, 578 (1967); Tschermaks Mineral. Petrog. Mitt. [3], *12*, 463 (1968).

121) Zachariasen, W. H., Plettinger, H. A., Marezio, M.: Acta Cryst. (Copenhagen) *16*, 1144 (1963).

122) Murdoch, J.: Am. Mineralogist *47*, 718 (1962).

123) Karashanow, N. A., Gabdshanow, Z. G.: Russ. J. Inorg. Chem. *11*, 747 (1966).

124) Aleksandrow, N. M., Ušakow, W. M.: Nachr. Leningrader Univ. *21*, Nr. 10, Ser. Physik, Chem. 2, 61 (1966).

125) Paton, F., McDonald, S. G. G.: Acta Cryst. (Copenhagen) *10*, 653 (1957).

126) Krogh-Moe, J.: Acta Cryst. (Copenhagen) *23*, 500 (1967).

127) Aschirow, A., Rumanowa, J. M., Below, N. W.: Ber. Akad. Wiss. UdSSR *143*, 331 (1962).

[128] Da-Nean Yeh: Science (China) *14*, 1086 (1965).
[129] Pennington, K. S., Petch, H. E.: J. Chem. Phys. *33*, 329 (1960).
[130] — — J. Chem. Phys. *36*, 2151 (1962).
[131] Lobanowa, W. W.: Ber. Akad. Wiss. UdSSR, *143*, 693 (1962).
[132] Clark, J. R., Erd, R. C.: Am. Mineralogist *48*, 930 (1963).
[133] Lehmann, H.-A., Kessler, G.: Z. Anorg. Allg. Chem. *354*, 30 (1967).
[134] Abdullajew, G. K., Mamedow, Ch. S.: Aserbaidshan Chem. J. *1965*, 101.
[135] — — Kristallografija, im Druck.
[136] Lehmann, H.-A., Papenfuss, H.-J.: Z. Anorg. Allg. Chem. *298*, 130 (1959).
[137] Takeuchi, Y.: Acta Cryst. (Copenhagen) *5*, 574 (1952).
[138] Berger, S. V.: Acta Chem. Scand. *3*, 660 (1949).
[139] Ito, J., Morimoto, N., Sadanaga, R.: Acta Cryst. (Copenhagen) *4*, 310 (1951).
[140] Kurkutowa, Je. N., Rumanowa, J. M., Below, N. W.: Ber. Akad. Wiss. UdSSR *164*, 90 (1965).
[141] Aschirow, A., Rumanowa, J. M., Below, N. W.: Ber. Akad. Wiss. UdSSR *147*, 1079 (1962).
[142] Kondratjewa, W. W.: Kristallografija *9*, 916 (1964).
[143] Lobanowa, W. W.: Ber. Akad. Wiss. UdSSR *135*, 173 (1960).
[144] Lehmann, H.-A., Zielfelder, A., Herzog, G.: Z. Anorg. Allg. Chem. *296*, 199 (1958).
[145] Schäfer, U. L.: Neues Jahrb. Mineral., Monatsh. *1968*, 75.
[146] Hart, P. B., Brown, C. S.: J. Inorg. Nucl. Chem. *24*, 1057 (1962).
[147] Osol, Ja. K., Wimba, S. G., Jevins, A. F.: Kristallographie (UdSSR) *9*, 32 (1964).
[148] Krawtschenko, W. B.: J. Strukturchem. (UdSSR) *4*, 271 (1963).
[149] Zeigan, D., Kutschabsky, L.: Monatsber. Deut. Akad. Wiss. Berlin *7*, 867 (1965).
[150] — Z. Chem. *7*, 241 (1967).
[151] Clark, J. R., Appleman, D. E., Christ, C. L.: J. Inorg. Nucl. Chem. *26*, 93 (1964).
[152] — Acta Cryst. (Copenhagen) *12*, 162 (1959).
[153] — Christ, C. L.: Z. Krist. *112*, 213 (1959).
[154] — — Z. Krist. *114*, 321 (1960).
[155] Moenke, H.: Naturwiss. *49*, 7 (1962).
[156] Christ, C. L., Clark, J. R., Evans, H. T.: Acta Cryst. (Copenhagen) *11*, 761 (1958).
[157] Holuj, F., Petch, H. E.: Can. J. Phys. *38*, 515 (1960).
[158] Clark, J. R., Appleman, D. E.: Acta Cryst. (Copenhagen) *15*, 207 (1962).
[159] Erd, R. C., McAllister, J. F., Vlisidis, A. C.: Am. Mineralogist *46*, 560 (1961).
[160] Lehmann, H.-A., Schaarschmidt, K., Günther, J.: Z. Anorg. Allg. Chem. *346*, 12 (1966).
[161] Christ, C. L., Clark, J. R.: Am. Mineralogist *45*, 230 (1960).
[162] Clark, J. R.: Am. Mineralogist *49*, 1548 (1964).
[163] Meixner, H., Moenke, H.: Kali und Steinsalz *3*, 228 (1961).
[164] Lehmann, H.-A., Herzog, G.: Z. Chem. *2*, 378 (1962).
[165] Allen, R. D., Kramer, H.: Am. Mineralogist *42*, 56 (1957).
[166] Weir, C. E., Schroeder, R. A.: J. Res. Nat. Bur. Std. *68A*, 465 (1964).
[167] Marezio, M., Plettinger, H. A., Zachariasen, W. H.: Acta Cryst. (Copenhagen) *16*, 390 (1963).
[168] Lehmann, H.-A., Jäger, H.: Z. Anorg. Allg. Chem. *326*, 31 (1963).
[169] Kutschabsky, L.: Z. Chem. *5*, 110 (1965).
[170] Switzer, G.: Am. Mineralogist *23*, 409 (1938).
[171] Braitsch, O.: Beitr. Mineral. Petrogr. *6*, 352 (1959).

172) Clark, J. R.: Z. Anorg. Allg. Chem. *331*, 348 (1964).

173) Jäger, H., Lehmann, H.-A.: Z. Anorg. Allg. Chem. *323*, 268 (1964).

174) Lehmann, H.-A., Kessler, G.: Z. Anorg. Allg. Chem. *360*, 267 (1968).

175) Gandymow, O., Rumanowa, J. M., Below, N. W.: Ber. Akad. Wiss. UdSSR *180*, 1216 (1968).

176) Clark, J. R., Christ, C. L.: Naturwiss. *55*, 648 (1968).

177) Braitsch, O.: Beitr. Mineral. Petrogr. *6*, 366 (1959).

178) Guertler, W.: Z. Anorg. Chem. *40*, 335 (1904).

179) Bartl, H., Schuckmann, W.: Neues Jahrb. Mineral. Monatsh. *1966*, 253.

180) Krogh-Moe, J.: Nature (London) *206*, 613 (1965).

181) Perloff, A., Block, S.: Acta Cryst. (Copenhagen) *20*, 274 (1966).

182) Berger, S. V.: Acta Chem. Scand. *7*, 611 (1953).

183) Wimba, S. G., Jevins, A. F., Osol, Ja. K.: Russ. J. Inorg. Chem. *3*, 325 (1958).

184) Kutschabsky, L.: Monatsber. Deut. Akad. Wiss. Berlin *9*, 512 (1967).

185) Osol, Ja. K., Jevins, A. F., Petch, L. Ja.: Nachr. Akad. Wiss. LettSSR. Ser. Chem. 3, 382 (1967).

186) Krawtschenko, W. B.: J. Strukturchem. [UdSSR] *4*, 768 (1963).

187) Lehmann, H.-A., Mühmel, K., Sun Dzui-Fang: Z. Anorg. Allg. Chem. *355*, 238 (1968).

188) Block, S., Perloff, A., Weir, C. E.: Acta Cryst. (Copenhagen) *17*, 314 (1964).

189) Mighell, A. D., Perloff, A., Block, S.: Acta Cryst. (Copenhagen) *20*, 819 (1966).

190) Block, S., Perloff, A.: Acta Cryst. (Copenhagen) *19*, 297 (1965).

191) Lehmann, H.-A., Sperschneider, K., Kessler, G.: Z. Anorg. Allg. Chem. *354*, 37 (1967).

192) Steele, W. C., Decius, J. C.: J. Chem. Phys. *25*, 1184 (1956).

193) Schuckmann, W.: Neues Jahrb. Mineral. Monatsh. *1968*, 80.

194) Schulze, G. E. R.: Z. Physik. Chem. *B 24*, 215 (1934).

195) White, J. G., Miller, A., Nielsen, R. E.: Acta Cryst. (Copenhagen) *19*, 1060 (1965).

196) Smith, P., Garcia-Blanco, S., Rivoir, L.: Z. Krist. *115*, 460 (1961).

197) Chang, C. H., Margrave, J. L.: Inorg. Chim. Acta (Padova) *1*, 378 (1967).

198) Ferrari, A., Coghi, L.: Gazz. Chim. ital. *71*, 129 (1941).

199) Lehmann, H.-A., Schmidt, W.: Naturwiss. *52*, 159 (1965).

200) Lonergan, G. A., Simpson, W. B.: Electrochim. Acta (London) *11*, 1495 (1966).

201) Peterson, R. C., Finkelstein, M., Ross, S. D.: J. Am. Soc. *81*, 3264 (1959).

202) Heller, G.: J. Inorg. Nucl. Chem. *30*, 2743 (1968).

203) Wlassowa, E. W., Valyashko, M. G.: Russ. J. Inorg. Chem. *11*, 822 (1966).

204) Valyashko, M. G., Wlassowa, E. W.: Jenaer Rundschau *14*, 3 (1969).

205) Hall, N. F., Sprinkle, M. R.: J. Am. Chem. Soc. *54*, 3469 (1932).

206) Bates, R. G., Pinching, G. D.: J. Am. Chem. Soc. *72*, 1393 (1950).

207) Kolthoff, J. M.: Z. Anorg. Allg. Chem. *115*, 168 (1921).

208) Tennyson, Ch.: Fortschr. Mineral. *41*, 64 (1963).

209) Zachariasen, W. H.: Acta Cryst. (Copenhagen) *16*, 380 (1963).

210) — Acta Cryst. (Copenhagen) *16*, 385 (1963).

211) Ihara, M., Krogh-Moe, J.: Acta Cryst. (Copenhagen) *20*, 132 (1966).

212) Burzloff, H.: Neues Jahrb. Mineral. Monatsh. *1967*, 157.

213) Hurlbut, C. S., Taylor, R. E.: Am. Mineralogist *22*, 1052 (1937).

214) Finney, J. J., Kumbasar, J., Clark, J. R.: Naturwiss. *56*, 33 (1969).

215) Dunicz, B. L.: Science (Washington) *153*, 737 (1966).

Eingegangen am 26. November 1969

ISBN 978-3-540-04821-3   ISBN 978-3-540-36201-2 (eBook)
DOI 10.1007/978-3-540-36201-2

Titel-Nr. 7723

SPRINGER-VERLAG
BERLIN·HEIDELBERG·NEW YORK

Fortschritte der chemischen Forschung

Herausgeber: A. Davison, Cambridge, MA; M. J. S. Dewar, Austin, TX; K. Hafner, Darmstadt; E. Heilbronner, Basel; U. Hofmann, Heidelberg; K. Niedenzu, Lexington, KY; Kl. Schäfer, Heidelberg; G. Wittig, Heidelberg

Schriftleitung: F. Boschke, Heidelberg

Band 14 / Heft 1
Inorganic and Analytical Chemistry

With 60 figures. 126 pages. 1970. Soft cover DM 38,— US $ 10.50

Contents: H. A. Bent, Localized Molecular Orbitals and Bonding in Inorganic Compounds. — W. D. Ehmann, Non-Destructive Techniques in Activation Analysis. — R. B. King, The Fragmentation of Transition Metal Organometallic Compounds in the Mass Spectrometer.

Band 14 / Heft 2
Technische Organische Fluor-Chemie

Mit 5 Abbildungen. 108 Seiten. 1970. Geheftet DM 34,— US $ 9.40

Contents: O. Scherer, Technische organische Fluorverbindungen.

Band 14 / Heft 3
Preparative Organic Chemistry

131 pages (including 59 pages in German). 1970. Soft cover DM 48,— US $ 13.20

Contents: S. Hünig und H. Hoch, Acylierung von Enaminen (160 Literaturzitate). — W. K. Musker, Nitrogen Ylids (183 References).

Band 14 / Heft 4
Carbohydrate Chemistry

With 8 figures. 211 pages. (including 126 pages in German) 1970. Soft cover DM 78,— US $ 21.50

With contributions by: H. Behre, J. S. Brimacombe, M. Černý, R. J. Ferrier, C.-P. Herold, A. Kraus, F. W. Lichtenthaler, H. Paulsen, H. Simon, J. Staněk.

In kritischen Übersichten werden in dieser Reihe Stand und Entwicklung aktueller chemischer Forschungsgebiete beschrieben. Sie wendet sich an alle Chemiker in Forschung und Industrie, die am Fortschritt ihrer Wissenschaft teilhaben wollen.

In der Regel werden nur Beiträge veröffentlicht, die ausdrücklich angefordert worden sind. Schriftleitung und Herausgeber sind aber für ergänzende Anregungen und Hinweise jederzeit dankbar. Manuskripte können in den „Fortschritten der chemischen Forschung" in Deutsch oder Englisch veröffentlicht werden.

Jedes Heft der Reihe ist auch einzeln käuflich.

This series presents critical reviews of the present position and future trends in modern chemical research. It is addressed to all research and industrial chemists who wish to keep abreast of advances in their subject.

As a rule, contributions are specially commissioned. The editors and publishers will, however, always be pleased to receive suggestions and supplementary information. Papers are accepted for "Topics in Current Chemistry" in either German or English.

Single issues may be purchased separately.